A BEGINNER'S BOOK
OF
TINCRAFT

Other Books by Lucy Sargent
Tincraft
Tincraft for Christmas

A BEGINNER'S BOOK OF TINCRAFT

Lucy Sargent

Designs and drawings by the author

Photographs by Bradbury/McCormick

DODD, MEAD & COMPANY · NEW YORK

Library of Congress Cataloging in Publication Data

Sargent, Lucy.
    A beginner's book of tincraft.

    Includes index.
    SUMMARY: Directions for making chimes, mo-
biles, ornaments, jewelry, and other decorative and
functional objects out of tin cans. Includes all the
techniques necessary for dealing with the medium.
    1. Tinsmithing—Juvenile literature. 2. Tin
cans—Juvenile literature. [1. Tinsmithing.
2. Tin cans. 3. Handicraft] I. Title.
TT266.S36   745.56   76-12506
ISBN 0-396-07354-9

Merci, Henri

CONTENTS

A BEGINNER'S BOOK
OF
TINCRAFT

Millicent Jones' First Tin Expressions. Ms. Jones' familiarity with flowers is evident in these lively portraits of chrysanthemums and roses from her garden, but who would believe she had never made anything from tin cans before?

1

WHY TIN CANS?

When ABC-TV came to my house to film a mini-documentary on tin-craft, the cheery young producer with his crew set up the camera and sound track in the living room, stood me in my accustomed corner by the desk, turned on the blinding klieg lights and said, "Now, tell me, what *is* tincraft?"

Well, I had to think a second. I'd been so busy just *doing* it over the last few years, filling up three books full as I went along, that I'd never once thought to ask myself that question.

Quickly, all the things I'd had such fun making flashed through my mind—as well as visions of others still in my head, to be made the minute I could get to them. Such an unbelievable wealth of possibilities, all so amusing—how could I possibly sum it up?

Can you imagine, then, how amazed I was to find that in reducing my craft to terms, I could only say, "Tincraft is making things out of tin cans." The statement was so pitifully inadequate, it seemed to hover before me in the glare of the lights, waiting to be rescued. I rushed on to add, "All *sorts* of things! *Beautiful* things!"

Still distressingly unconvincing, but time was of the essence and I had to go on with the show. So I pointed to my collection of cans, talked about tools, and demonstrated how easily one could make an elaborate-looking sculpture by taking it one simple step at a time.

Only much later did I realize why I'd gotten into difficulty with my definition. I'd instinctively felt that it should, somehow, sum up *all* aspects of the craft—the facts, of course, but the scope, too, and *particularly the pleasures*.

What I hadn't fully realized was that the pleasures of a craft are derived only from doing it—not from reading, or from listening, or from watching —just from doing. So, you see, I was really expecting too much of my definition, and I hope that you will come along with me to *do* tincrafting and thereby discover the pleasures for yourself.

Tin cans are made from tinned sheet steel called tinplate, and we use them because they are available to everyone everywhere, which tinplate is not. Tinplate must be purchased from a distributor at considerable expense in large boxes of no less than ninety sheets, and most of us would never use that much, at least not as beginners. Furthermore, it makes no sense for us to invest a lot of money in tinplate when we can get a perfectly good substitute for nothing. After all, when you make something out of nothing, you are 100 per cent ahead!

You're ahead with tin cans in other ways too, because they come already "turned, locked, and grooved," as the first patent machine for working tinplate was described in 1805, and they come in an astonishing variety of shapes and sizes. The American Can Company alone makes some twenty thousand different kinds of cans, if you can conceive of it, and that means you have a head start when making something from cans, whether it's a small windchime of lids or a large chandelier from the whole can: *all of the measuring and most of the shaping and soldering has been done for you by the manufacturer.*

Still another advantage of the ready-made can is the ribbing, which the manufacturer, for some reason even he doesn't know, calls *beading*. Probably you've noticed the concentric circles in the lids, and the variety of ribs in the sides, which are put there to make the can more rigid. For us craftsmen, these beadings are not only decorative, but useful, in that they serve as guidelines. Because they are there to crimp or cut along, we don't have to use a compass to mark the lid when crimping a candlecup, or use a scriber on the sides when cutting a strip of tin for the scroll to hold the candlecup.

More often than not you will hammer the beading flat to make the metal more flexible and easier to cut, but the markings will remain visible, so you still have guidelines to go by and still have their ordered, mathe-

14

matical design. Sometimes, both the lids and sides of cans are perfectly *plain*, and that's good too, because you will often need a piece of pure, unblemished metal.

Then, there's the color. Most people never think to look *inside*, to see the amazing array of gold-lacquered linings that cans possess, ranging in hue from palest champagne to deepest ruddy gold put there by the manufacturer to protect the can from the aggressive chemicals in the food. Nor do most people look *under* the label to see that some cans are mirror-bright, others softly frosted, and still others marked in faint white diamonds. Some are even gold on the *outside*, as well as the inside, which means they are not only twice as splendid but twice as durable, because they are protected on *both* sides by the lacquer.

So you see, all things considered, you are working, at no cost, with metal that is visually just as beautiful as gold and silver, and that has already been shaped to your advantage. True, the tinned sheet steel that tin cans are made of is less durable and malleable than gold and silver, but all the same it is quite strong enough to make objects of lasting beauty: Christmas ornaments, candleholders, wall sculptures. And while tinplate is not malleable enough to make a deep bowl, it has enough give in it to make a perfectly satisfactory ash tray or nut dish.

Later, I will introduce you to my favorite cans, so that you can keep an eye out for them, but first I'd like to tell you how they came to be invented.

As you know, nothing in life happens all by itself or all at once. Evolution is a slow process. So I don't suppose it is too surprising that even though *tin* had been discovered as early as 4000 B.C., the tin *can* was not invented until A.D. 1810. In the interim, people had learned a lot about the properties of tin: that it was soft, so soft it could be hammered into paper-thin sheets known as *tinfoil*, but that when just a little of it was combined with another soft metal, namely copper, a remarkably *strong* alloy was formed called *bronze*. This discovery was so important to man, you remember, that he named an Age for it.

People also learned that a large amount of tin, combined with a very little lead, made another most serviceable alloy called *pewter*; that tin combined with chlorine made a fixative for dyeing cloth; that tin salts

The Empress Maria Theresa's Tin Tomb. Because of tin's resistance to corrosion, Maria Theresa and her husband, Francis I, continue to disport themselves atop their tomb as brightly as they did when they were buried in it in 1780.

could advantageously weight silk, could fireproof materials, could preserve wood and paint, even under water. Of particular interest to me is the fact that tin is essential in type metal, without which this book would not have been printed. Of vital interest to all of us is that tin is indispensable in *solders* and *bearing* metals, without which *nothing* would run—*no cars, trains, or planes, telephones, radios, or TV's!*

But the characteristic of tin which is of immediate interest to us, as tin-can craftsmen, is its capacity to resist moisture and bacteria, and to protect whatever it covers. People discovered that a thin coating of tin on iron would keep the iron from rusting.

Just about the time that tinned iron plate was being produced in what anyone could describe as a continuous process, Napoleon's armies were marching across Europe "on their stomachs," and they desperately needed portable supplies of unspoiled meat and vegetables. So the French government offered a prize to the person who could invent a practical way of preserving food, and in 1809, Nicholas Appert won that prize with his method of steaming it in tightly corked jars. One year later, an English merchant named Peter Durand won a similar prize in his country for successfully storing food in "tin canisters." *And there we are!*

The canning industry was thus inaugurated, and, as you might expect, the process has by now been altogether revolutionized. The old heavy hand-hammered iron plate, hand-dipped in molten tin and hand-polished, has been replaced by 20,000-foot coils of mirror-bright steel, formed in monstrous rolling mills by a continuous cold-reduction process, moving at sixty miles an hour and passing on through an electroplating tank to emerge with a coating of tin 75 *millionths* of an inch thick! It seems incredible that such a thin coating could transform perishable steel into durable tinplate, but, in fact, it does.

As you start cutting up cans, you will notice that some are heavier and harder than others. That is because there are so many thousands of products to be canned that the steel industry must supply tinplate of four different chemical specifications, of ten different gauges or degrees of thinness, of seven different temper rollings, of nine different tincoat thicknesses, and of fifteen basic enamels. Sometimes the inside tin coating is heavier than the outside and that's when you'll find those diamonds I mentioned earlier. They indicate to the can manufacturer which side should be inside. In any case, all these differences work to your advantage. The light, limber Campbell's Soup cans make the best Christmas tree ornaments and jewelry, for example, and the heavy V-8 Juice cans make stunning star sculptures and handsome, durable candlescrolls.

You might be amazed and amused to know that in 1974 there were *87.1 billion* cans manufactured in the United States. That comes, if you can believe it, to over four hundred per person, and most people think they don't use any at all! Twenty-one per cent of these cans were made of aluminum, about which I am not so enthusiastic. Aluminum lacks the crispness and resilience of tinned steel. I prefer to work with the regular, three-piece, side-seamed cans. They are a heavier weight than the seamless

My Favorite Cans. The large, shallow can on the left is used commercially for fillets of sole; the tall one standing in it transports salad to the A & P; the ribbed can on the right contained Idahoan Instant Dehydrated Mashed Potatoes.

cans and are usually paper-labeled rather than painted. So seldom have I been able to capitalize on the art work printed on a can, and it's such a nuisance to remove, that I rarely use a lithographed can unless, for reasons of size or weight, I need that particular piece of metal.

Now, for my favorites!

Among the retail cans, look out for the following, always remembering that *because they are subject to an evolving can technology, they may change:*

Campbell's Soup, regular size, heads the list, especially Beef Broth, Black Bean, Chicken Gumbo*, Chili Beef*, Consommé, Green Pea, Hot Dog Bean*, Old Fashioned Tomato Rice*, Pepper Pot, Scotch Broth,

18

Tomato Beef Noodle O's*, Tomato Bisque*, Vegetable and Beef Stockpot*, and Vegetarian Vegetable*. There must be *something* on that list you want to eat!

These cans are made of a light, limber grade of tinplate that is a joy to work with. They all have ravishing rich-gold linings, ideal for reproductions of ancient jewelry in particular, and ornaments in general. The ones with an asterisk after them are marked on the outside under the label in diamonds, which inspired me and I hope will inspire you to make the Diamond Belt, Collar, and Curtain on pages 93-98. Tomato Soup also has diamonds on the outside, but does not have the lovely gold lining.

I have just learned that only the Campbell plant at Camden, New Jersey, which services New England, uses *straight-walled* cans. All the other plants are using beaded (ribbed) cans which are not so all-purpose. Campbell's offers no explanation for this inconsistency.

V-8 Juice, large size, also made by Campbell's, has always been my next favorite. It was, without question, the handsomest juice can on the market until recently when Campbell's changed over to a *khaki-colored* lacquer on the lid, making it less than perfect. The sides, however, still have their rich-gold linings and are made of the same heavy grade of metal, ideal for candlescrolls and large star sculptures.

Coffee-can lids, any size, are all-silver and make super Frosty Stars, coasters, and crosses. Unfortunately, the sides are usually lithographed.

Crisco and Spry, any size, provide an alternative source of *un*lithographed, *un*beaded, all-silver cans, good for all sorts of projects.

Idahoan Instant Mashed Potatoes and Durkee's French-Fried Potato Sticks come in crisp, all-silver, *beaded* cans, measuring as much as six inches across, terrific for stars and mirrors.

Ideal Dog Food cans have handsome *green*-gold interiors which make for highly visible Christmas tree ornaments. You may have noticed, incidentally, that most pet food comes in cans with drab, gray-lacquered linings, but cheer up: if you're painting the ornament an *opaque* color, it doesn't matter what color the lining is.

Friskies Buffet used to have bright gold interiors and maybe they still do where you live; they seem to vary from place to place.

B & B Mushrooms, any size, have the deepest *red*-gold linings I have found. They are straight-walled.

Bumble Bee Brand Tuna, large size, is ideal for the Sunflower Sconce on page 135.

Cream-Style Corn, any brand, has consistently soft-gold linings with fine, clustered beading.

Dolmas, (Greek stuffed grape leaves), have richly lacquered linings and come in straight-walled cans up to six inches in diameter, excellent for the Mexican Sun Mirror on page 141.

Green Giant Asparagus has an interesting *tortoise-shelled* gold lining.

Sweet Potato and Yam cans, any brand, are rich-gold, straight-walled, and generally all-purpose.

Van Camp's White Hominy has an interestingly *mottled*-gold lining, especially suitable for Near Eastern and Oriental pendants.

In addition, there is an impressive selection of huge, commercial cans, readily obtainable from the delicatessen departments of most supermarkets. They can also be procured, on occasion, from restaurants, fish markets, and specialty shops in large condominium complexes:

Atalanta Polish Ham, twelve-pound size, because of its so-called pear shape, is the choicest can for sculpturing pineapples. The European metal is soft and pewter-y in hue and the configuration of the beading lends itself perfectly to the shape of both the fruit and the leaves.

Krakus Polish Ham, eleven-pound size, now seems to be coming through in the oblong *pullman* can, whose sides provide an excellent source of plain, unbeaded metal. I made the large Cruising Fish Mobile on page 109 from this can.

Salads of all kinds come to the delicatessen department of my A & P in the most spectacular, tall, circular cans I have ever seen, *blazing gold, inside and out*, ten inches wide and eight or twelve inches high! Fresh blueberries and apple turnover mix come to our local restaurant in similar cans. The Celebration Chandelier on page 146 was made from such a can.

Fillets of Sole are sold commercially in New York's Fulton Fish Market in large, shallow, all-silver cans, which I have used for most of the Superstructures in this book. Supermarkets with first-rate fish departments would be your best source for these.

Soybean oil and Cashew nuts come to our local restaurant in tall, *square*, all-silver, five-gallon cans, which are even more ideal than the oblong pullman can for any project where you need a large piece of metal, as for a picture frame. In fact, each side of these square cans has beading resembling a picture frame.

You will discover other beauties for yourself, without a doubt. Just keep your eyes open and your hands out!

2

DON'T YOU CUT YOURSELF?

Let's face it! "Don't you cut yourself?" is the first question that enters people's minds when they learn you cut up tin cans. They look at your fingers curiously to see if they're covered with scars and are amazed to find they're not. When my nails are broken and my knuckles bruised, it's from weeding the garden or raking leaves, not from cutting up cans. In fact, almost everything else I do in the way of chores, like washing the dishes or polishing the brass, is harder on my hands than tincraft. So when somebody asks me if I cut myself, I always laugh and say "Never!" Then, of course, I have to add, "Well . . . *hardly* ever." Because sometimes I get so interested in developing an idea that I forget about the laws of nature! But I've never gotten more than a scratch and the most I've done is put a Band-Aid on it. Usually, I just lick it and forget it.

You will learn very quickly how to handle tinplate, always picking it up lightly instead of grabbing it, and holding it firmly so that it doesn't slip. Of course, you avoid running your fingers along the raw edge, just as you would along the blade of a knife, but in no time at all you will know so exactly how to proceed, you'll be using your fingers to shape an ornament almost as often as you do the pliers.

It's the *little short stiff cuts* you have to watch out for. For instance, when you are opening up a can to get a piece of metal to work with, you have to cut down alongside the rigid seam. This is hard going; the snips seem to get stuck and your instinct is to back out and try again. But when you re-enter, your snips can make a little hangnail as you start cutting again and this hangnail will prick you if you're not careful. Once you know that, though, you will always be on the lookout, and as you get more used to opening up cans, you'll learn how to avoid making those hangnails in the first place by keeping the snips deep in the cut. One thing you will be relieved to learn is that opening up the can is really the hardest part of tincraft. If you can do that—and *of course* you can!— you can do anything.

You may want to wear gloves to start with, just so you won't have to worry while you're gaining familiarity with the action of the snips and the *re*action of the metal, but once over your strangeness, you will probably find gloves more in the way than anything else. I never wear them—except to church!

Always work in a good light, daylight preferably, because there is less glare on the metal. You can make out the markings more easily and see where you're going. The southwest light in my living room seems ideal. That's partly because, with floor to ceiling windows, there is plenty of it. But also because there is an agreeable warmth to it, in comparison with the cold light of a north window or the dazzling light of a south.

Work near a wastebasket or over your hammering board to catch the clippings as they fall. Because you usually have just hammered the piece of metal you are going to use, and are sitting there with the board on your lap anyway, the board makes an especially convenient surface to work over. Better height, too, than your desk or table. Which reminds me that one of the neat things about this hobby is that you can do it anywhere. You don't have to have a studio or even a workbench—just a toolbox, table, chair, and wastebasket. Except for the spray painting and the soldering, I've done all the projects for three books right at my desk in the living room, and I could even have done the soldering there, if the torch had struck me as an acceptable piece of living-room sculpture.

But to get back to the cutting, do not brush the snippings off the board into the wastebasket *with your hands*. Either pick the pieces up lightly and place them in the basket, or tip them off the board into the basket. That way, no scraps get stuck, either in your fingers or in the rug.

Allow plenty of time even if you don't think you'll need it. Time is your only investment, after all—except for a good pair of snips—so don't be in a hurry. Start with a lid and play with it. Approach it musingly with a "What if I were to . . . ?" or a "How would it look if . . . ?" attitude.

23

Seven-year-old Christopher Knox. Chris is cutting up cans for the first time and see what he's made in an hour!

Make friends with your medium, and the first thing you know, you'll find the metal giving *you* ideas, and *together*, you will come up with something absolutely unique. I've seen that happen so often that I honestly now believe that if a hundred people were given identical lids to experiment with, they would each come up with a different design. It's amazing how individuality surfaces in this craft. Perhaps it's because the metal is free and that, in turn, frees the imagination. In any case, I have learned that no matter how explicit the instructions may be, no one seems to reproduce a design *exactly*. Somewhere along the way, his own personality takes over and influences his interpretation. And I say, *"Vive la différence!"*

One last word about nicking yourself. When I wrote *Tincraft for Christmas*, my editor said, "If, as you say, anyone from eight to eighty can do tincraft, we'll have to prove it by getting an eight-year-old to make something."

So, I called up my next-door neighbor, Chris Knox, who was only seven actually, and asked him if he'd mind being a guinea pig. An hour later, without ever having done *anything* of the sort before, Chris had made a mobile of bells and stars. What's more, he'd never even thought about cutting himself!

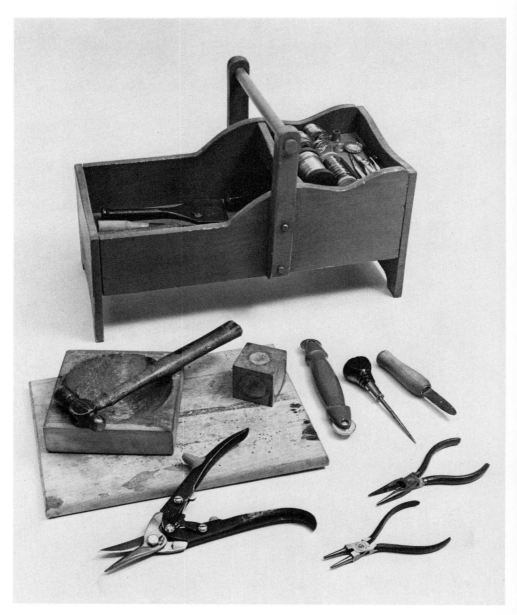

Tools for Tincrafting. *Front Row:* Klenk Aviation Snips, round-nosed pliers, long-nosed pliers. *Middle Row:* Hardwood hammering board, chopping bowl, ball-peen hammer, wood dapping block, screen installation tool, awl, oyster opener. *Back Row:* A small cobbler's bench is not essential, but is big enough to hold all the tools and materials needed for tincrafting. The handle can even be used for curving candlescrolls.

3

TREAT YOURSELF
TO SUPERIOR SNIPS

Most of the tools you need for tincraft are ordinary ones you probably have around the house, such as a hammer, a screwdriver, pliers, chisels, and such. But the one tool you will rely on more than any other and which will make all the difference between delight and despair are the snips. You really should treat yourself to a good pair! When you consider that your raw material is free, superior snips are not an inordinate indulgence, and as a beginner, you need the right tool as much as anybody.

My favorites, and those used by many metalworkers in the trade, are the Klenk Aviation Snips. They are the *only* ones I can recommend without reservation. Unlike old-fashioned tin snips, they cut both straight *and in circles*. Their blades are so finely milled they can sheer off a hair's breadth of metal if necessary. Their handles are comfortably pistol-gripped to fit the palm, rather than looped (which can blister the fingers), or wing-spread (which overtires the hand). And because of their double-cam action, they cut through steel as if it were butter.

I have two pairs, one plain-edged and one serrated. The serrated snips have thirty-six tiny teeth along the blades which produce a brilliant coruscation on the edge of the metal. This makes them especially desirable for Christmas ornaments, and inasmuch as Christmas inspires us to make things more than any other season of the year, it is a real temptation to have a pair.

On the other hand, if you feel that, to begin with, you should invest in only one pair, the plain-edged snips are perhaps more all-purpose. Our ancestors had only plain-edged snips, and in reproducing any authentic Early American design, they are the ones to use. They are also best for any utilitarian object like a letterholder; or anything primitive, like the Mexican Bird; or for most jewelry where you usually need a smooth edge.

Because Klenk snips are used largely by the trade, you will not find them in the average hardware store and will have to send for them mail order (see Sources of Supply). Specify left or right hand, since you have the choice. Should your snips ever need it, Klenk will sharpen them for you for a small fee, but unless you turn professional and do a great deal of cutting, these snips should last you a lifetime and give you endless satisfaction.

4

OTHER ESSENTIAL TOOLS

Besides the snips, there are a number of other tools you will use constantly.

Can Opener. The one I have used is the wall-type Can-O-Mat Series DL 245 S2, made by Rival Manufacturing, which removes rims neatly. This not only saves you the unpleasant task of cutting them off with the snips, but gives you attractive rings to design with. Even though Rival has now phased out the wall-type model, there are still a good number of them left in the hands of distributors. In any case, there are other brands which will remove rims, both automated and manual, including Rival's own Click 'n Clean and Sears Insta-clean. Should you need to replace your opener, take a can along with you to the hardware store to test the ones on display. The clerk will tell you it's impossible to remove rims, but he'll let you try anyway, and you'll go home triumphant, leaving him much amazed.

Ball-peen Hammer. This round-headed, flat-ended hammer is indispensable. The flat end smooths the metal, making it easier to cut and less hazardous to handle. The round end cups the metal, giving it dimension, and enriches the visual texture. Just be sure if purchasing a ball peen that the head is perfectly round, *no point at all*. Jewelers' supply houses are the most reliable source (see Sources of Supply). Six (or eight) ounces is a comfortable weight.

Hammering Board. A *hard*wood board about a foot square (one solid piece, not laminated), with a straight-edge and true corners, is ideal for making awl holes, hammering metal smooth, and folding over edges of candlescrolls or desk-pad corners.

Pliers. You will need two kinds: *Long-nosed* pliers whose blades are round on the outside, *but flat on the inside*; and *round-nosed* pliers whose blades are *completely round*. And their blades *must meet the full length* or they won't grip the metal effectively. Utica manufactures the best I know of.

Awl. A sharp, pointed tool used mostly for making holes, but also to scribe lines and "punch" ornamental designs on metal. You could use a nail, but the handle of an awl is much easier to hit than the head of a nail, and you will need it so often, it's worth the small investment.

5

OPTIONAL SUPPLIES

Then, there are a number of other tools and materials you will find convenient.

Architect's Linen. A transparent, durable, *glazed cloth*, ideal for patterns which you plan to use many times, because it is much less likely to tear when you pull it off. You will have to mail order it unless your local art store happens to stock it, but a yard should last practically forever.

Chisels. Both *wood* and *cold* chisels are useful for making ornamental impressions on tinplate, as are assorted sizes of screwdrivers.

Compass. An inexpensive compass will do nicely for scribing around the sides of a can you want to cut down or divide in two, as with the Mexican Sun Mirror.

Kitchen Chopping Bowl. Used with the ball-peen hammer, to curve *all* of a surface. **Dapping Block** (preferably one deeply hollowed out from a block of wood as shown) is used to curve only a *portion*.

Knives. There are two kitchen knives you may find use for: a short, stiff, boning knife for cutting a can in two; and an oyster opener for making slots in things like pierced lanterns.

Nail Sets. Use these for "punching" ornamental designs in metal like those in the Iced Drink Coasters.

Paper Punch. When hit with a hammer sufficiently hard to penetrate the metal, this will make neat round eyes for birds and fish.

Steel Wool. Ordinary kitchen soap pads are fine for shining up a can before setting to work, but to finish off, you may want to add a little extra sheen with the very fine 0000 steel wool.

31

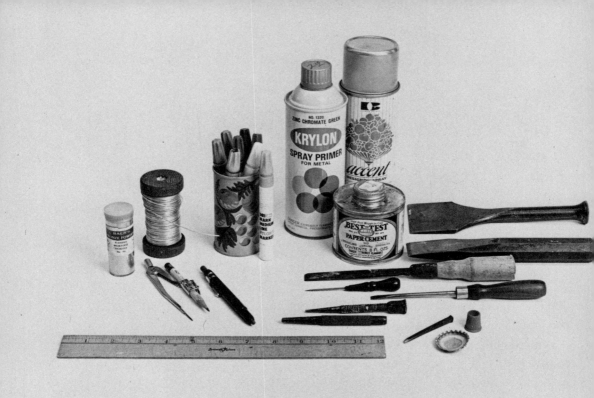

Miscellaneous Tools and Materials. *Front Row:* Ruler, compass, grease pencil, two nail sets, stiletto, chisel, screwdriver, sawed-off nail, beer bottle cap, toothpaste cap. *Back Row:* Baer's bronzing powder, Nu-Gold wire, permanent felt markers, Krylon's Zinc Chromate Green Spray Paint, Illinois Bronze Tôle Red Spray Paint, rubber cement, and two cold chisels.

Upholsterer's Tacks. These brass tacks harmonize with the metal and make excellent ornamental focal points.

Wire. You will have use for all weights of wire from very fine hair wire to No. 18 heavy galvanized wire. Brass welding rods also come in various weights and have the advantage of being straight-stemmed rather than coiled, like most wire which makes them especially good for flower stems. For jewelry, you should mail order a spool of Nu-Gold (copper/brass) wire; No. 18 is the best all-purpose weight.

Glues. *Epoxy 220.* Even though soldering is the classic method of bonding metals, you needn't feel guilty about using epoxy; you can be sure our ancestors would have used it if they'd had it. It is ideal for cementing candlesockets to candlecups, and candlecups to candlescrolls.

32

Always score or sand the surfaces to be bonded. On the other hand, epoxy is "forever" and you must be sure your design is right.

Otherwise, the best all-purpose glue I've found is *Weldit Cement*; it grips well, especially if you have scored the area first, but you *can* pry it off if necessary. *Rubber cement* is the *only* glue to use for applying patterns to tinplate because it can be peeled off, leaving a clean surface. Always apply it to the can, *not* to the pattern. For gluing mirrors, you should beg some *mastic* from the man who cuts them for you; other glues attack the silver coating.

Paints. The quickest way to apply opaque color to metal is to spray it on; *Krylon, Illinois Bronze, and automotive spray paints* cover well and dry fast.

However, if you want to draw *designs* on metal, and want the tin to show through, use either *permanent felt markers* or, for a more durable finish, transparent artist's oils, diluted in a lot of marine spar varnish.

6

TECHNIQUES

The can should be squeaky clean to start with, the glue under the label scraped off with a paring knife, or if sticky, with paint remover; the price mark removed with mineral spirits or turpentine; the silver sides polished with steel wool; and the luscious, lacquered lining washed with a soft, soapy cloth or sponge to avoid marring its perfection.

The Effect of Snips on Tinplate

Let me tell you a secret: the most complicated-looking designs in this book are among the easiest to make, and this is because of the effect snips have on tinplate. When you cut a strip off the *right*-hand side of a piece of tinplate, the action of the snips encourages the strip to curl *down under*. When you cut a strip off the *left*-hand side, the snips encourage it to curl *up above*. *You* are cutting *straight ahead*, but the strip is curling either *down* or *up*. The narrower the strip, the tighter the curl. The crisper the metal, the tighter the curl. But, in order for the strip to curl at all, the metal must be *smooth* and *flat*, and this brings us to the importance of hammering.

The Importance of Hammering

If the can is beaded (ribbed) instead of straight-walled, you will find it difficult to cut along the beading unless you have hammered it flat first. It *can* be done, but it's tricky. So also with the beading in the lid when cutting a curly star: you must hammer the metal until it is smooth and supple before attempting to cut curls. This means that with most projects you will hammer before, during, and after. Wear earmuffs if you must, place the board on your lap or on the rug to help absorb the sound, but hammer!

Hammering not only facilitates cutting, but it also adds texture and value to the surface of the metal. This is especially appealing when making anything from nature, like flowers or birds, but in every instance, hammering reduces the mirror brightness of the sheet metal and gives it a subtle dimension and a hand-worked appearance. Much as we craftsmen appreciate the pristine perfection of the can just as it comes from the manufacturer, we do not want to be forever reminded that it once was a can. We want to transform this utilitarian object into an *objet d'art*. Hammering helps to do this.

You will notice that when you hammer, even with the flat end of the ball peen, the metal will *curve up* to it, so it is possible to achieve dimension in your design simply by hammering it from *one* side. If, however,

Scallop Shell and Revolutionary Ash Tray. Two examples of shallow bowls formed by beating with the round end of the ball-peen hammer on a flat board.

you want the piece perfectly flat, you must hammer it from *both* sides.

By using the round end of the ball peen, you get an even more marked dimension and textural effect, particularly suitable for flowers and fruits. When used on a flat board, the ball peen will produce a shallow cup; if used in a chopping bowl, a deeper cup—even a bowl if you persist; and if used with a dapping block, an interior portion of a design can be raised, leaving the rest untouched.

Removing Rims with a Can Opener

Let's hope you have an opener which will remove rims, because, as you can see from the Frosty Star Mobile on page 39, rims can be used decoratively when removed neatly, whereas they are totally disfigured when cut off with snips. To determine whether or not your can opener will remove rims, turn the can sidewise and slip the rim under the wheel of the opener *with the seam just this side of the wheel.* Then, when you crank the handle the wheel should come down full force on the seam and cut through it easily, and you *ought* to be able to continue on around, severing

Removing Rims with a Wall Can Opener. Some wall can openers will remove rims when the cans are slipped in sidewise under the wheel.

the rim neatly. If the wheel should miss the seam, however, don't worry. Just crank around until you come to it again, remove the can from the opener, and bend the rim back and forth till it snaps off.

You might sometime want to make a simple sconce using the whole end of a can, rim and all. In that case, do not remove the bottom lid, but proceed merely as if to remove the rim, and the entire end will come off intact.

Opening Up the Can to Obtain the Metal in the Sides

To open up a can to get a piece of metal to work with you must first remove the bottom lid. Then, if you are right-handed, you will cut along the *left* side of the stiff seam, so that the more limber body of the can will give a little and ease your struggle.

Bracing the snips against your stomach and grasping the far end of the can with your helping hand, *hold all firmly together* and slug your way down through the metal. Keep the snips *deep* in the cut at all times and *never withdraw*, no matter what! If the nuts on the snips get in the way, waggle them to the right. Do not be tempted to pull out and start over, for that will only cause the snips to sliver up little hangnails as you start cutting again. And don't try coming up from the far end to meet the cut, for that also creates a jagged point where it fails to make a smooth connection.

Whatever the situation, don't be discouraged! You will soon get used to slugging your way through cans and think nothing of it. Comfort yourself with the thought that this is the single most awkward thing to do in tincraft; everything else is easy!

Once you have cut through the can, trim off the seam. Then, if it is a beaded can, cautiously but firmly, pull the sides apart as far as you easily can. Place it *convex side up* on your hammering board and step on it. Next, hammer all around the edges until they're smooth; then hammer the beaded surface until it is nicely flat and limber.

If the can is straight-walled, you will need only to hammer around the edges. Then place the piece *concave side up* on your knee and draw it back and forth, over and down, to flatten it out.

37

Using a Boning Knife to Obtain the Metal in the Sides

Some cans are too large for a wall can opener to remove the rim or the whole end of the can, but there is another way to get at the metal in the sides which may seem unorthodox, but which works surprisingly well. Take a sharp, short, boning knife. Plunge it into the side of the can next to the seam (you may have to tap it with the hammer to get it started) and cut around the can as if it were a loaf of Boston brown bread. You'll be amazed to discover how easily and controllably the knife moves through the metal and how smooth the raw edge is. You can also divide a can into two *un*equal sections, layering the smaller on top of the larger, as with the Mexican Sun Mirror on page 141.

Cutting Down Cans

Every now and again a can is too tall for your purposes and you will want to cut it down. With your compass opened to the height you want, hold it against the outside of the can with the pencil point touching the bottom rim.

Keeping your eye on the pencil to make sure it is always touching the rim, move the compass around the can, pressing on the stylus to scribe a line on the side of the can. You use the compass in this unconventional way because most tinplate is too crisp to accept pencil marks, but it can always be scratched by the stylus.

Once the line is defined, cut down along the left side of the seam as you always do to open up a can, and *curve over gently* to join it. Continue cutting carefully along the scribed line, pressing the can against your knee for a firm purchase, and bending back the loose metal to facilitate cutting.

Fringing Tinplate

One of the easiest ways to gain a decorative effect on the edge of a design is to make straight cuts into it, ⅟₁₆″ or ⅛″ apart. As you can see from the Sunflower Sconce on page 136, this is especially effective when done with *serrated* snips.

Frosty Star Mobile, illustrating the techniques of fringing and ornamenting rims.

Ornamenting Rims

When rims are removed by a can opener, a small strip from the side of the can is included with it. You have the choice of cutting it off with the snips and filing the rim smooth, or of ornamenting the rim by cutting straight through the strip to the rim, as if fringing it. The action of the snips will turn the metal in each little cut, achieving a highly decorative effect, as you can see from the photograph above.

39

Pennsylvania Punched Tin Box with favorite heart and weeping willow motifs. *Courtesy, The Henry Francis du Pont Winterthur Museum*

Punching Tinplate

We usually associate the term "punching" with making a hole in something, but Early American tinsmiths pulled their punches on purpose so as not to make holes but merely to raise a decorative design on the surface as in the Punched Box above.

Piercing Tinplate

Piercing, however, does penetrate the metal. The most popular example of this in our country is the pierced barn lantern, which for generations we have been incorrectly calling the Paul Revere lantern. Near Eastern in origin, it was brought to Spain by the Moors and from thence through Portugal to New Bedford, Massachusetts, in the early seventeenth century.

We can approximate the round and oval piercing with an awl and oyster opener, but constructing the lantern requires the skill of a professional tinsmith.

Striking Awl Holes

To make a neat awl hole use a *hard*wood board. Strike it first from the front, then from the back; and hammer it flat from the back.

Dividing a Lid into Equal Sections

You will need to divide a lid into equal sections constantly, and it may cheer you to know there's a way to do it *by eye*—no measuring! The first few times, you might want to put a dot in the center, or trace around a dime, but after a while you won't need to. You will simply hold the lid right up in front of you, looking at it squarely, and cut straight up from six o'clock to within a dime's worth of the center. Then turn the lid around so the cut is up top at twelve o'clock and come right up from six again. Turn these cuts to nine and three o'clock and come up from six again.

Continue to divide each section in half to make as many sections as you need, always facing the section squarely so that you can judge the midpoint accurately.

Dividing the Sides of Cans into Equal Sections

You can also divide the sides of cans into equal sections *without measuring*. First you must cut a piece of paper to *butt around the can perfectly within the rims*. Fold the paper in half, creasing it sharply. Do this three or four times or until the sections are about ¾″ wide.

Brush the can with rubber cement. Smooth the paper around the can, butting it at the seam, and cut down along the creases to the bottom rim, thus dividing it into equal sections.

Sometimes you may want to remove the top rim before smoothing the paper on. Other times, for stars or suns, you may want to draw straight

41

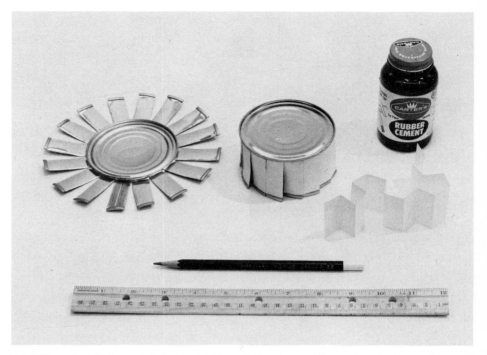

Dividing the Sides of Cans into Equal Sections. The creases in a piece of paper, cut to butt around the can and folded in half several times, provide guidelines to cut along, thus dividing the sides of the can into equal sections. Rubber cement allows the paper to be peeled off easily. Always apply it to the can, not the paper.

or wavy lines in each of the sections *before* you smooth the paper on, and then *cut along those lines instead of the creases.*

Typing paper is a good size for small cans; shelf paper for larger cans; and coarse tissue paper for enormous cans; but *not* newspaper, which absorbs the glue and sticks to the can.

Dividing the Sides of Cans into *Unequal* Sections

Cut a piece of paper to butt around the can between the rims, as usual, and fold it *almost* in half, but leave approximately ¾″ of the paper *extending beyond the double thickness.* Continue to fold the *double thickness* in half until the sections are ¾″ wide.

You may have to decrease the width of that extra section if the circumference of the can does not divide itself naturally into ¾″ sections, but once you understand the principle involved here, you can easily make that adjustment.

Dividing a Rim into Three Equal Intervals

Cut a narrow strip of paper to butt around the rim. Fold the ends in, over each other, to form three equal sections. Mark on the rim the points where the creases come.

Tracing Patterns

If you don't plan to use the pattern more than once, any translucent paper will do, but if you plan to use it repeatedly, use architect's linen (see Sources of Supply).

Should the instructions read: *Place on fold of linen*, trace the design on the folded linen and glue the fold together with rubber cement *before cutting it out*. You will then get a perfect mirror image.

Cutting Out Patterned Designs

Any pattern will adhere best to a smoothly hammered piece of tinplate. Apply it with rubber cement and hold the piece with your helping hand in such a way that the pressure of your fingers will not move the pattern out of position as you cut.

If you are right-handed, cut *counter*clockwise around the pattern to keep it on the "up" side of the snips. When you come to a corner with such a tight reverse curve that you cannot swoop around it—like the neck of the Gingerbread Man on page 91, for example—cut right into the corner where his head joins his shoulder. *Withdraw* the snips and start cutting again farther along his arm, *bypassing* that awkward curve. Continue on around into the next tight corner, bypass, and so on. Once you have him all cut out, hammer the corners smooth and clean them up by cutting back into them from the *opposite* direction.

Making Eyes with Awl or Paper Punch

An enlarged awl hole makes a perfectly good eye for a bird or fish. Center the spot for the eye over the hole in your spool of Nu-Gold (or any large spool of thread), and hammer the awl through the metal up to the hilt. If the hole isn't large enough to suit your purpose, insert your round-nosed pliers into it and waggle them around to enlarge it. Repeat this from the back and hammer the hole flat from the back.

A flawless hole can be made with a paper punch. I happen to have two kinds. The larger, *flat*-tipped punch must be hit with the hammer in order to stamp out the circle, but the smaller one has an *angled* tip which penetrates the metal merely with the pressure of the hand. The one limitation of the paper punch is the length of its shaft: you can only work ¾″ in from the edge of the design.

Finding the Balance Point in Mobiles and Hanging Ornaments

To determine the balance point in the arm of a mobile, tie the string, cord, or fishline around the dowel, wire, or rod, and move it along until the construction balances. Glue it in place with epoxy.

To find the balance point in a hanging ornament, suspend it by a thread which has been *Scotch-taped* to the point where you think the hole should be. Shift the Scotch tape to right or left as required, and strike an awl hole at the proper point.

Making Links

In the trade, what you and I call *links* are called *jump rings*. You can buy them so cheaply, from the same place you would order the wire to make them from, that you may decide to do so, but I shall give you the procedure in any case.

To make *round* jump rings, for situations where there will be no stress to speak of, wind Nu-Gold or No. 20 galvanized wire around a pencil (or knitting needle of equal size) at least as many times as you need rings.

Slip the coil off the pencil and cut through it with your snips. The rings will fall off as you cut, all ready to be used.

To make *oval* jump rings, wind the wire around *two*, small, *metal* needles, which, together, equal the size of a pencil. Slip the coil off, and carefully cut through the *center of one side* of the oval coil. Then, when you link your pieces together, the pressure will come on the *ends* of the ovals, rather than on the opening in the side.

Of course, jump rings really should be soldered together, but that's tricky. There's no good substitute for solder, but you might try epoxy.

Making Scrolls and Candlecups

A scroll is a curved strip of metal used either to support the candlecup of a sconce, or suspend a windchime from the top of a door. To give it the necessary strength and resilience, its edges must be folded over.

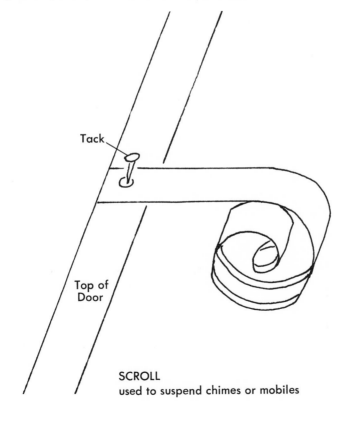

Tack

Top of
Door

SCROLL
used to suspend chimes or mobiles

Hammering Board

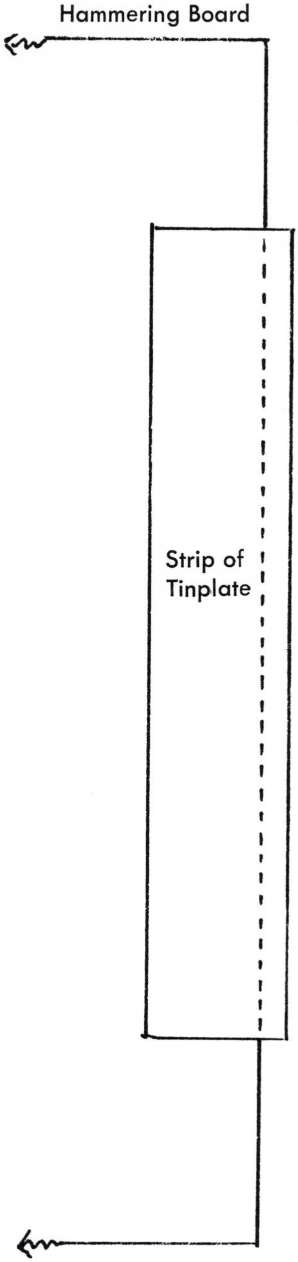

Lay strip of tinplate along
straight-edge of hammering
board with 3/16" overhang

Hold firmly and hammer overhang
down at right angles to side of board

Stand strip on edge and continue
to hammer overhang down

Lay flat and *crease well*. Repeat
on other side of strip.

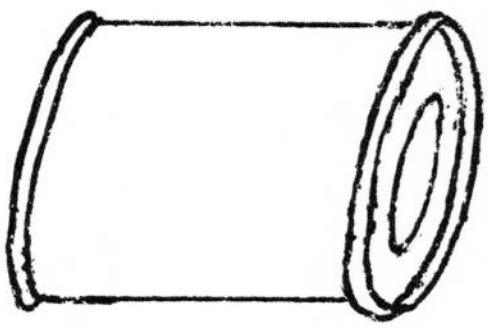

Strip of
Tinplate

Remove whole end of can and beat with
round end of ball peen to form candlecup

Curve scroll on dowel held in vise,
striking just to right of center

From the sides of a hammered, beaded can, cut a strip of tinplate 1⅜″ x 10″ and place it along the straight-edge of your hammering board with a ³⁄₁₆″ overhang.

Holding the strip firmly, hammer down the overhang at right angles. Stand the piece on edge and continue to tap the overhang down. Lay the piece down and hammer the fold flat, creasing it well. Repeat along the other side of the strip.

Cut a V notch in one end of the strip and hammer it around a dowel or short length of pipe, held in a vise, until you've achieved an easy graceful curve. Or, if you have no pipe or vise, choose one of the simpler, square-angled scrolls pictured here.

I usually make candlecups from the rimmed lid, pried off a small frozen juice container, but you can use the whole end of a soup can just as well. Beat it with the round end of the ball-peen hammer on your hammering board, kitchen chopping bowl, or dapping block until it is nicely cupped.

If you plan to use a taper instead of a votive candle, you will need to form a candlesocket from a strip of tinplate 1″ x 3″. Hammer this around

SCROLL ALTERNATIVES

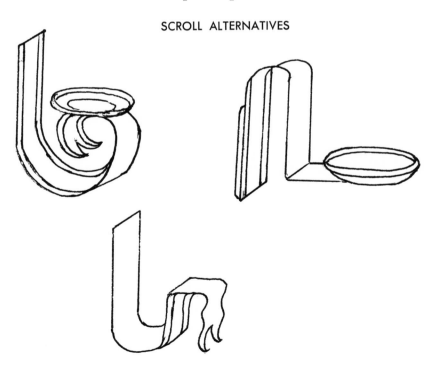

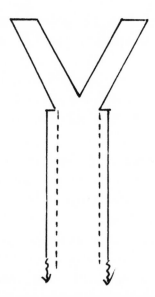

STAR OF DAVID FORKED SCROLL

a broom handle or pipe to overlap ⅛". Glue the overlap with epoxy. When dry, seat the candlesocket in a little pool of epoxy in the center of the candlecup.

Wait to glue the candlecup to the scroll until after you have glued the scroll to the sconce back. Then hang the sconce on the wall. Apply epoxy to the crest of the scroll and the *bottom* of the candlecup. When they are both sufficiently tacky, set the candlecup in position, keeping an eye on it to make sure it doesn't slip to one side.

Crimping

By 1805, American tinsmiths had mechanical crimpers, but you can do a creditable job by hand just with pliers. Long-nosed pliers give a more square-cut look to the crimping than round-nosed, so take your choice.

To crimp a candlecup, hammer the lid flat. If you are right-handed, grasp the edge of the lid with the pliers to a depth of approximately ⅜" and twist your wrist to right and left, forming a little valley and hill. Move the pliers into the little hill on the right and twist right and left again, forming another valley and hill. Keep moving into the next little hill on

48

the right, working counterclockwise around the lid. You may have to go around twice to make the crimping stand up properly. Some of the early sconces have perfectly vertical crimping, but the flared seems more graceful to me.

Soldering

Soldering is now taught in many elementary schools, starting in the fourth grade, which gives you an idea of how easy it *has* to be.

The principle is simple: two similar metals, when *perfectly clean* and *heated sufficiently* to melt the proper solder, will *bond* together. If you have the right equipment, soldering is quick and easy.

You will need:

Asbestos Mat. Western Auto carries thin rolls for less than a dollar. Tack it onto your hammering board.

Propane torch. A Bernz-O-Matic is the one I use.

Dunton's Tinner's Fluid. Actually a weak solution of hydrochloric acid. Western Auto sells this in a bottle with a dispenser top, which is ideal. Otherwise, it must be applied to the hot metal with a natural bristle brush.

Kester's 50/50 Solid Core Solder (no acid or rosin core).

To show you how easy soldering is, let's go through the motions required by the Satellite Mobile on page 52.

Lay a lid, silver side up, on the asbestos mat.

Place the wire on it.

Heat the area where they touch until the wire begins to turn orange.

Immediately squirt some tinner's fluid on it. It will sputter as it cleans the metal.

Place a snippet of solder alongside the wire.

Apply heat until the solder melts. Wherever the metal has been sufficiently cleaned by the acid, the solder will run around like quicksilver.

Watch it cool before moving it.

Wash off the acid immediately to stop its corrosive action.

(Note: Where you don't need the design of the concentric circles, hammering both the lid and the wire *flat* makes for a better bond.)

7

LET'S LOOK AT A LID

You have just fed the cat, let's say, or had a cup of soup, and there you are with the lid of the can—a shining gold and silver disc, impressed with perfect concentric circles, and, except for the ripple around the edge, quite pristine and flawless—when viewed objectively. In fact, if you were a Man from Mars and didn't know its purpose was purely utilitarian, you might think the lid was an ornament. Even on our planet, as recently as the fourth century B.C., the Scythians and the Egyptians wore similar circles of gold sewn on their clothing or linked together in collars around their necks. Perhaps you and I, conditioned as we are, would hesitate to sew lids on our shirts and skirts, but we can certainly enjoy the sound and sparkle of them in mobiles and windchimes.

Wafer Wind Chime

Once you have drunk enough juice or eaten enough spaghetti to collect the little lids, this wind chime can be put together in a matter of minutes. Nine lids are the bare minimum; sixteen are super!

The wafers need only be tapped around the edge to minimize the ripples. Strike an awl hole near the edge of each and link them together, one above the other, on three (or four) single strands; on *individual* strands, they get tangled up as they sway in the wind. If you're worried about weathering, spray them with Krylon Crystal Clear.

At Christmastime, use red wool instead of fishing line; a red dowel instead of a brass tube; and sprigs of green to celebrate the season!

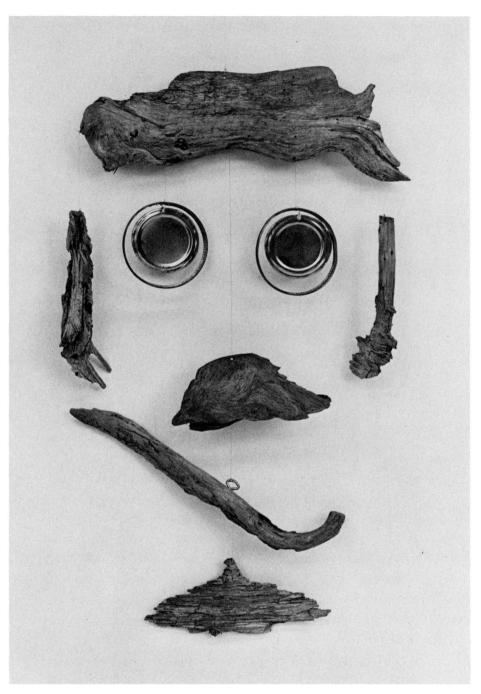

Little lids make big eyes in this driftwood Santa Claus.

Wafer Wind Chime

Satellite Mobile

Paint it red, white, and blue—for the Fourth of July!

You will need a 4″ Styrofoam ball and little cans of spray paint from the dime store, a coil of No. 18 gauge galvanized wire from the hardware store, a heavy wire coat hanger, and about thirty-six lids.

For variety, I used two sizes of lids: twenty-four 2¾″ dog food lids, and twelve 2″ frozen juice lids. I soldered them onto 8″, 12″, and 16″ wires (see the diagram). If you prefer, you could swirl the tips of the wires in epoxy and lay them on the lids, but, of course, that takes additional drying time. In any case, solder or glue before painting.

When it comes to painting, you will have one complication: *all the wires must be painted white* so they won't disappear from view in the night light. This means you will have to cover the red and blue lids carefully when you spray the wires the contrasting color.

52

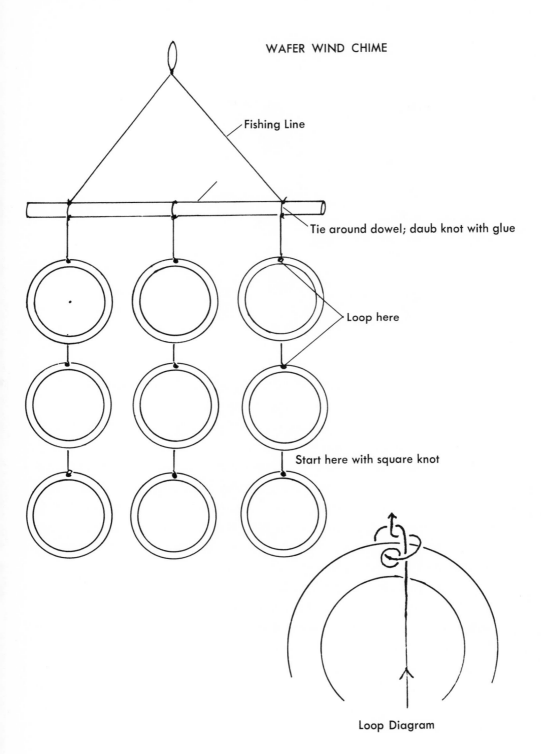

WAFER WIND CHIME

Fishing Line

Tie around dowel; daub knot with glue

Loop here

Start here with square knot

Loop Diagram

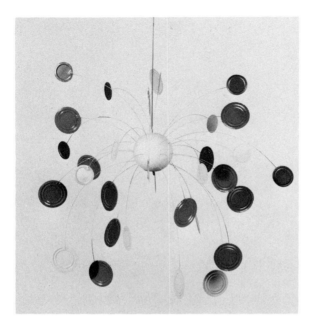

Satellite Mobile

Also, be sure to let each coat of paint dry thoroughly, so the lids will not stick to the newspaper they are resting on.

I suggest you start by spraying four of the 2″ lids and eight of the 2¾″ lids, *and* their wires, white.

Then spray eight of the 2″ lids blue, and sixteen of the 2¾″ lids red. When thoroughly dry, cover them carefully and spray their stems white.

In between sprayings, you can work on straightening out the coat hanger with the long-nosed pliers, leaving the hook for hanging at the top and nipping off the other end two feet below. Be sure the rod is perfectly straight and vertical before you impale the Styrofoam ball upon it.

To reinforce the bottom of the ball where the rod emerges, beat an extra 2″ lid with the round end of the ball-peen hammer to cup it up. Strike an awl hole in the center and spray it red. When it is dry, slip it on the wire, and with the pliers, form a loop on the end to hold it in position.

The only rule to follow in spacing the satellites is one of balance. For the mobile to hang properly, you must work in groups of four, starting at the bottom of the ball with four long-stemmed reds, sticking them into the ball at the four points of the compass: one right, one left, one front,

54

SATELLITE MOBILE

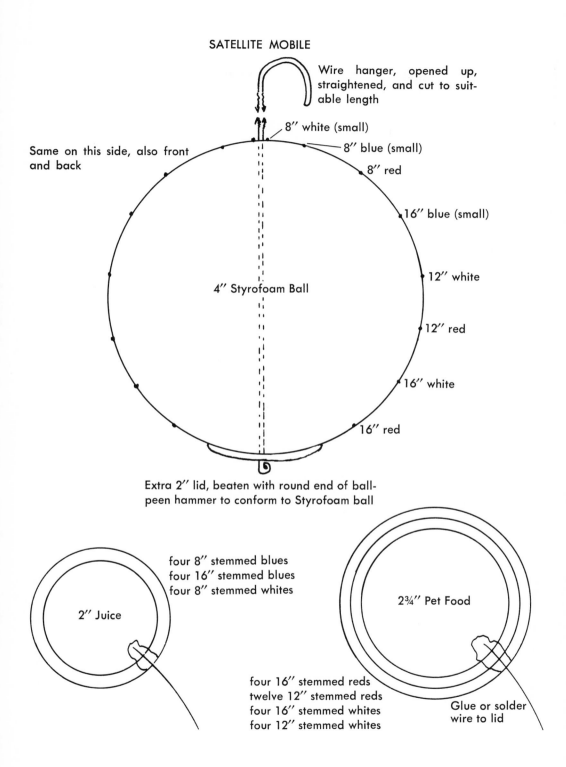

Wire hanger, opened up, straightened, and cut to suitable length

8″ white (small)

8″ blue (small)

8″ red

16″ blue (small)

Same on this side, also front and back

12″ white

12″ red

4″ Styrofoam Ball

16″ white

16″ red

Extra 2″ lid, beaten with round end of ball-peen hammer to conform to Styrofoam ball

2″ Juice

four 8″ stemmed blues
four 16″ stemmed blues
four 8″ stemmed whites

2¾″ Pet Food

four 16″ stemmed reds
twelve 12″ stemmed reds
four 16″ stemmed whites
four 12″ stemmed whites

Glue or solder wire to lid

and one back. They don't need to hang down evenly, just so long as they are opposite each other. In fact, it's more interesting if they are a little uneven.

The next set of four long-stemmed lids should be white, and be stuck into the ball about an inch above the set of reds. Then a set of *medium-stemmed* reds should be stuck in the ball an inch above the whites, and so on. Your eye will tell you what to put where, but even if you find you're wrong, don't worry about destroying the ball. It's amazing how durable Styrofoam is and how it conceals one's trials and errors.

I gave my mobile to the small boy whose Mum saved the lids for me and we were all quite pleased with it. There's something so stirring about red, white, and blue!

Iced Drink Coasters

By tapping a lid, from front and back, with screwdrivers, chisels, and assorted nail sets, it is possible to ornament it without cutting into it. This technique is called *punching*, in contradistinction to *piercing*, which actually penetrates the metal.

Our ancestors also practiced a form of engraving called *wriggling*, which involved many short strokes of the chisel and achieved a remarkably refined effect. Unfortunately, our twentieth century tinplate is too smooth and hard for wriggling, and almost too tough for piercing with anything but the sharp-pointed awl and oyster opener. But it *can* be punched, and all that is required is a hammer, punches, and patience!

The coasters you see photographed here show three stages of development, the simplest of them requiring two sizes of screwdrivers and a nail set struck from the back. The second joins the flowers in the first with scallops of awl marks struck from the back. The third adds an extra set of scallops from the back and a circle of awl marks from the front, inside the decorated edge.

You will notice that the center is not touched, or the concentric circle around it, so that the glass can sit securely upon it. If you were to decorate the center as well, however, you would wind up with something resembling a Near Eastern or Tibetan pendant, and that's something to think about!

ICED DRINK COASTERS

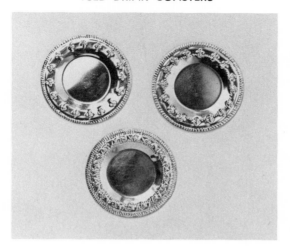

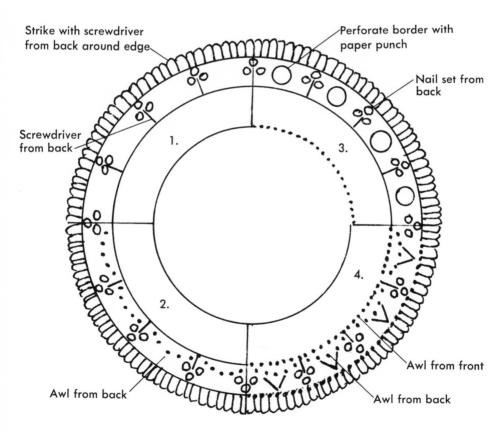

Strike with screwdriver from back around edge

Perforate border with paper punch

Nail set from back

Screwdriver from back

1.

3.

4.

2.

Awl from front

Awl from back

Awl from back

Four Alternative Designs

Great Gear

Great Gear

This striking ornament might take you all of five minutes to make. As you can see, it involves nothing but straight cuts in from the edge of the lid, with every other tab bent down.

To be sure you get an *even* number of tabs, look at the instructions under Techniques on Dividing a Lid into Equal Sections, and make the cuts around the edge in the order given there.

When you're ready to put the upholsterer's tack in the middle, make a hole for it with the awl to avoid hammering your fingers. Clip the point off the tack, fill the head with Weldit Cement, and place it in position. You may want to put another clipped tack on the back, if the ornament is to be seen from both sides.

If you think your gear is handsome enough to wear, you will need a chain. Ordinary plumber's brass or anodized aluminum chains from the hardware store are suitable and readily accessible, but only a little less expensive than the real thing mail ordered from a jeweler's supply house (see Sources of Supply).

The silhouette of these gears is so dramatic that if you have a picture window, or a sliding glass door, you might want to make a curtain of them, come Christmas!

Sparkler Star

Shining lids make super stars! This little sparkler with its notched ends is the easiest of all to make.

SPARKLER STAR

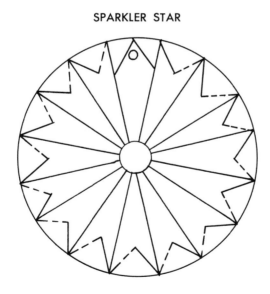

Divide lid into sixteen equal sections. Notch the end of each section *except* the one from which it will hang; cut that to a point, and strike awl hole.

Hammer the lid of a Campbell's Soup can perfectly smooth and divide it into sixteen equal sections (see Techniques), cutting up to within a dime's worth of the center. Then notch the ends of each section *except the one from which it will hang*. Cut that to a point and do it first or you'll forget it! Strike an awl hole in it for hanging.

You can paint the silver side of the star with Magic Markers and glue a rhinestone in the center if you like. You can also push one section back and pull the next one forward, or twist them first left and then right. I usually leave the little points curling, rather than hammering them flat, because they look so much more sparkly that way, but I have to admit they're more prickly too!

Sparkler Star Mobile

This mobile hangs year-round in my daughter's coral-colored bedroom and sparkles just as brightly now as it did eight years ago when it was first made. The sections in the stars are painted alternately shocking pink and orange, to match the Decorette Ribbon which encircles the rim above. But the mobile can be very charming in just the natural gold and silver of the can with gold and silver Decorette Ribbon to match.

For the frame you will need a rim 6″ in diameter (three-pound coffee or Idahoan Instant Mashed Potatoes or Crisco); an 18″ length of lacy Decorette Ribbon to glue onto it; and 5′ of gold tinsel Christmas cord for suspending it and ornamenting it with bows.

Cut three 9″ strands of tinsel cord and tie them to the rim at equal intervals (see Dividing a Rim into Three Equal Intervals). Trim the knots and daub them with glue. Pick up the tips of the strands and bind them together with hair wire about 1½″ down, making a loop for hanging as you do it.

Cut two 12″ lengths of gold cord for bows to be placed on either side of the "neck" and tie them in place with another 6″ length of cord. Suspend the frame from a hook where it will be handy for you to work on. (I have an almost invisible fishline and hook hanging from a beam over my desk in the living room.)

Next, make nine Sparkler Stars according to the directions given above

Sparkler Star Mobile

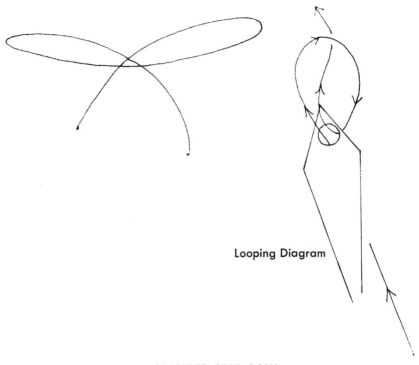

Looping Diagram

SPARKLER STAR BOW

and ornament them any way you wish. Cut three 9″ and three 13″ lengths of *fine* gold thread (bought by the spool in the dime store). Tie a star on the end of each of the 9″ strands and attach them to the frame in the *middle* of each interval so that the rhinestone in each star hangs at a point 8″ below the rim. Trim the knots and daub with glue.

Tie three more stars on the ends of the 13″ strands; loop the cord through the three remaining stars at a point 6″ higher up, fastening the tip of the strands to the rim where the tinsel cords are attached.

The rhinestones in the stars should hang at points 5″, 8″, and 11″ below the rim. Make any adjustments, trim the knots, and daub with glue.

Apply a fine line of glue to the rim and gently press the lacy ribbon into place. With hair wire, fasten bows made from 9″ lengths of tinsel cord at the key points on the rim. For a little extra emphasis and finish, dip the tips of the bows in glitter.

Swiss Starlets

Swiss Starlets

Apparently someone in Switzerland sat down on a cold winter's evening in plenty of time before Christmas and experimented with a lid just as you and I have been doing, saying to himself, "What if I were to . . ." We can be grateful to Mrs. Orvis Yingling for sharing this anonymous craftsman's stellar experiments with us.

You'll notice that the starlet on the lower left (A) is like the Sparkler, except that it has only eight, instead of sixteen, sections. *But*, this Swiss had a bright idea! For, in addition to notching the ends, he thought to cut a narrow strip along *one side of each section* and curl it up tight with

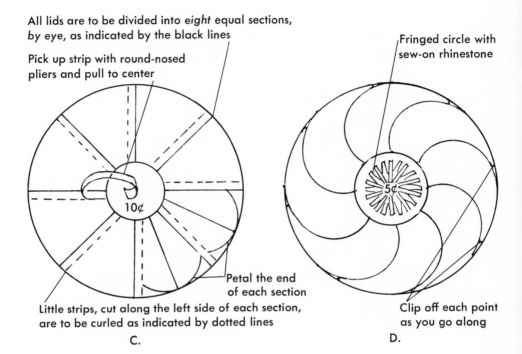

All lids are to be divided into *eight* equal sections, by eye, as indicated by the black lines

Pick up strip with round-nosed pliers and pull to center

Fringed circle with sew-on rhinestone

10¢

5¢

Petal the end of each section

Little strips, cut along the left side of each section, are to be curled as indicated by dotted lines

C.

Clip off each point as you go along

D.

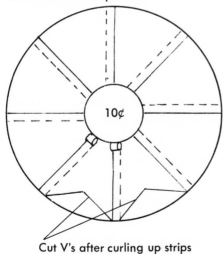

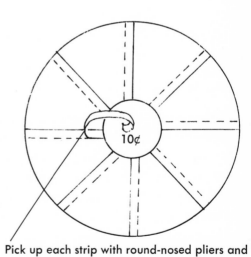

Curl up each strip tightly with round-nosed pliers

10¢

10¢

Cut V's after curling up strips

A.

Pick up each strip with round-nosed pliers and pull to center to form a whorl

B.

round-nosed pliers. By this very simple expedient, he created an instant flower-form.

In Starlets B and C, we see variations of the first, except that the narrow strip, rather than being rolled up tight, has been picked up with the round-nosed pliers and pulled to the center of the star, where, with each successive strip similarly tucked down, a whorl is formed. The only difference between the second and third starlets is the treatment of the *ends* of each section which have either been left *un*cut, or been curved like a petal.

The fourth starlet on the upper right, however, is a complete departure from the others, with the lid divided into *whorled* sections, like a pinwheel, and the ends of those sections have been curled with the round-nosed pliers to form little trumpets. You see, as the whorls were cut, *the tip of the next whorl curled down on the right under the snips.* This gave our Swiss the idea of grasping that curl with the round-nosed pliers and continuing its curve to form a trumpet. Very simple, but ingenious, don't you think?

Let me make a few suggestions to help you with this fourth starlet (D). After tracing around a nickel placed in the center of the lid, mark off, by eye, the four points of the compass, both on the outer edge of the lid and on the *inner* circle. Then mark four points *midway* between them, thus dividing both circles into eight sections. Draw arcs from the points on the perimeter, *not* to the corresponding point on the inner circle, but *to the next point above it*, leaping up, as it were.

As you cut these whorls, *clip off the point on the upcoming* section, so you won't get snagged by it. This advice may be hard to comprehend in the abstract, but when you start cutting, you'll see what I mean.

You will find the whorled starlet very pretty just as it is, without any trumpets, and you may decide to ornament the center with a fringed circle and rhinestone, and let it go at that. But if you want to form the trumpets, *turn the starlet over* and, grasping the curl with the *broad base* of the round-nosed pliers, continue the curve to form a trumpet. This will not be easy because you will be working against an *opposing* curve, but with a little practice, you'll get the knack of it.

Make a tiny loop on the tip of one trumpet for hanging.

Incidentally, these ornaments were made with Campbell's Soup lids and are therefore quite small. If you always have a huge tree, you might want to increase the size of the lid a little; in any case, keep all your ornaments to scale.

Swedish Star and Snowflake

If these ornaments look familiar to you, it's because you've seen them many times, carved out of wood. Craftsmen in Sweden have made a happy living doing just that for generations. And, happily for *us*, the lid of a can lends itself ideally to the design. It's so easy to make!

Hammer the lid until smooth. Cut a strip of cardboard or tin as long

Snowflake and Swedish Star

66

SWEDISH STAR AND SNOWFLAKE

For the star, cut a strip of tin the diameter of the can and suitably wide for its length. Place it across the center of the lid and mark along both sides with permanent felt marker.

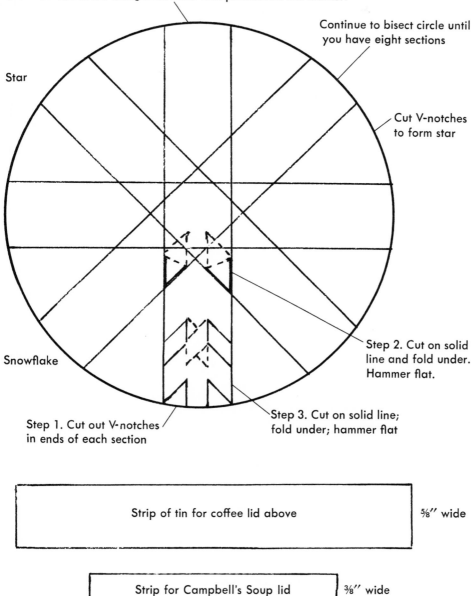

Continue to bisect circle until you have eight sections

Star

Cut V-notches to form star

Snowflake

Step 2. Cut on solid line and fold under. Hammer flat.

Step 1. Cut out V-notches in ends of each section

Step 3. Cut on solid line; fold under; hammer flat

Strip of tin for coffee lid above   ⅝″ wide

Strip for Campbell's Soup lid   ⅜″ wide

as the diameter of the lid and of suitable width, say ⅜″ for a soup can and ⅝″ for a coffee-can lid.

Lay the strip across the center of the lid, dividing it in half. Mark along each side of the strip with a fine felt-tipped pen.

Turn the lid at right angles and divide it again, thus forming a cross. Do this twice more, forming *eight* sections. Cut V-notches between the sections and there you are with a star! Spray it red if you like.

For the snowflake, make short parallel cuts in the ends, base, and sides of each section as shown. If you do this in the order given (i.e., Steps 1, 2, 3), it will all go along quickly; it is always much more efficient to do all of one kind of operation at once. Spray your snowflake snow-white!

Frosty Star

This is the most incredibly resplendent, coruscated collection of curls obtainable and, amazingly, it is created by *cutting straight ahead with serrated snips.* If you've read the section in Techniques called The Effect of Snips on Tinplate, you will know that when you cut a narrow strip of

Frosty Star
68

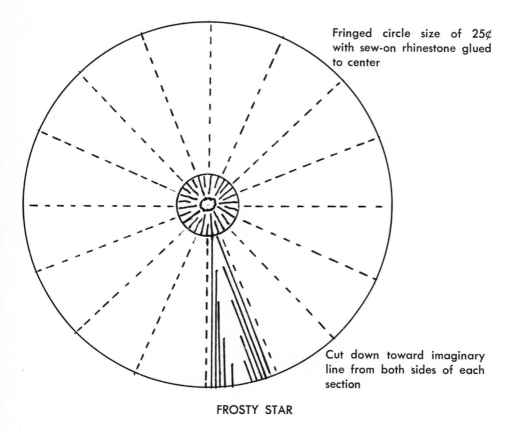

Fringed circle size of 25¢ with sew-on rhinestone glued to center

Cut down toward imaginary line from both sides of each section

FROSTY STAR

metal, it will curl *down* under your snips *on the right* and *up* above them *on the left.* Here you are cutting *straight ahead* and getting credit for having manipulated all those little curls, when it's really the tin and the snips that are doing it.

Besides the serrated snips, which you need for the crunchy coruscation, you must use an *all-silver* lid because a gold lining will diminish the brilliance. All coffee-can lids are ideal, as are the Idahoan Instant Mashed Potato can lids; and the greatest and most spectacular of all Frosty Stars are made with the huge commercial fish cans that flounder are packed in, measuring 12" across!

Whatever size lid you select, hammer it perfectly smooth, leaving no ridges of any kind, and divide it into sixteen equal sections (see Techniques). Then, if you feel you need to, draw a line down the center of each section (otherwise, imagine such a line) and cut narrow strips down *almost*

69

to it, first on the right side and then on the left. Occasionally a little strip won't curl enough to suit you, so just gently push it along in the same spirally curvy fashion the others have taken.

To crown the glory of it all, fringe a little circle of scrap tin, made by drawing around a quarter, and glue a sew-on type rhinestone in the center of it.

Frilly Star

As perhaps you can see, the Frilly Star is made just like the Frosty except that ten sections have been left *un*cut, providing a pleasant contrast to the frills at each side. To gain dimension, you can angle the plain sections.

Frilly Star

Mexican Star

The Mexicans use a much softer tinplate for their craft than the steel sheet available to us in cans, but that is at least partly because they produce the plate expressly for craft purposes, rather than for canning. As

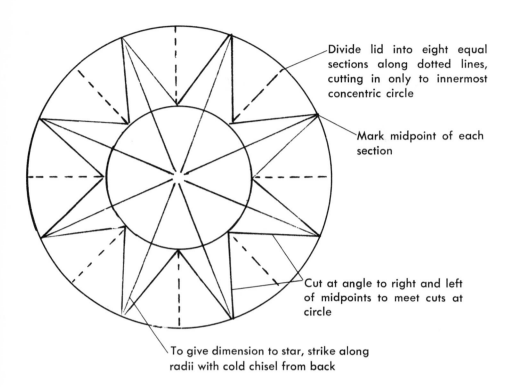

Divide lid into eight equal sections along dotted lines, cutting in only to innermost concentric circle

Mark midpoint of each section

Cut at angle to right and left of midpoints to meet cuts at circle

To give dimension to star, strike along radii with cold chisel from back

MEXICAN STAR

TRUE MALTESE CROSS

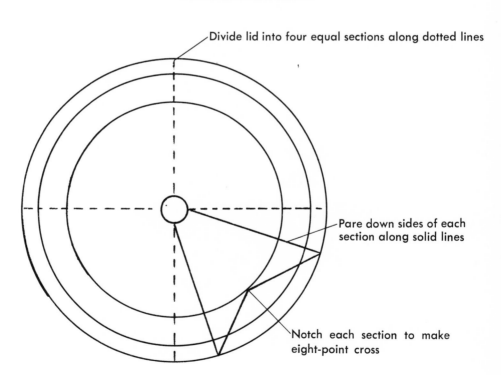

Divide lid into four equal sections along dotted lines

Pare down sides of each section along solid lines

Notch each section to make eight-point cross

a result, they can emboss (or press designs on) the metal easily by scribing around a template (metal pattern), thus delineating the graceful feathers of a bird or the veins in leaves. We, on the other hand, can make only a faint impression on our tougher metal with an awl, and perhaps a little more distinct impression with the hammer and chisel, but nothing to compare with the effective markings of the Mexicans. However, the hammer and cold chisel combine to give a nice dimension to this Mexican Star, made from the lid of a one-pound coffee can.

Hammer the lid well so that all the concentric circles are perfectly smooth. Divide it into eight equal sections, cutting only as far as the innermost concentric circle.

On the edge of the lid, mark the center point of each section with a felt-tipped pen and cut from those points *right and left at angles* to meet the cuts at the circle. With a broad cold chisel, hammer ribs in the rays, pivoting around the center point, marked with felt pen, from the back. Place the chisel accurately and strike it smartly.

True Maltese Cross

A cross of eight points is the symbol of St. John of Jerusalem and the most popular cross in heraldry.

Mark a ¼" dot in the center of a one-pound coffee-can lid and divide the lid into four equal sections, cutting right up to the dot.

Measure off points ½" in from the sides of each section, along the outer edge, and pare down the sections as shown in the diagram. Place dots in the center of the *third* concentric circle of each section and cut a broad V-notch to it from both corners. Nip off the points; hammer the cross from the back; and strike an awl hole for hanging.

Cross Pendant

It almost goes without saying that the lid you select for a pendant must be flawless, lustrous, and immaculate. The regular-size Campbell's Soup can provides the choicest source of such lids, being a most wearable 2½" in diameter and, more often than not, the color of spun gold. You can

73

Top: **Small Cross Pendant.** *Bottom:* **Pinwheel Pendant.**

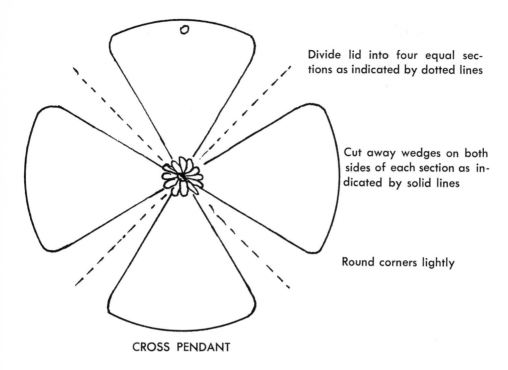

Divide lid into four equal sections as indicated by dotted lines

Cut away wedges on both sides of each section as indicated by solid lines

Round corners lightly

CROSS PENDANT

still wear a 2⅞" pendant comfortably, but this large 3¼" modified Maltese Cross requires a dramatic black velvet gown and the stature of a Wagnerian soprano!

As always, hammer the lid well and divide it into four equal sections (see Dividing a Lid into Equal Sections). Cut down the sides of each section as shown in the diagram and lightly round the corners. Ornament the focal point with an upholsterer's tack or anything suitable from your sewing basket. The chains on the crosses photographed here come from the hardware store; they are brass and anodized aluminum, matching the different shades of gold in the lids. You will notice that the larger cross is suspended by *two* links, the better to keep it from flipping over.

Pinwheel Pendant

Like the Cross Pendant, this Pinwheel Pendant is made by dividing the lid of a lustrous Campbell's Soup can into *four* sections. Cut a curve on one side of each section. Round the remaining corner lightly and beat

Divide lid into four equal sections as indicated by dotted lines

Cut curves on one side of each section as indicated by solid lines

Nip point off corners

Beat each section gently from back with round end of ball-peen hammer, *except along the straight-edge*

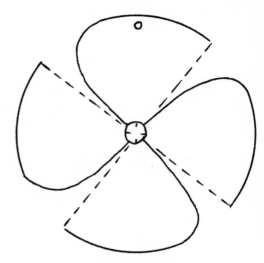

PINWHEEL PENDANT

the ornament from the back with the round end of the ball-peen hammer, *except along the straight-edge.*

Ornament the focal point as you wish. The sew-on type rhinestone is deeper than the flat-backed rhinestones you find in multicolored packets in the dime store, and has more distinction. You might want to shop around for small, sparkling flat-backed buttons for a change, and order a proper chain from a jeweler's supply house (see Sources of Supply).

Dazzling Zinnia Pin or Pendant

The inspiration for this zinnia came from a Tiffany ad in *The New York Times.* The original was just a zinnia with nothing but petals, but I happened to have the stone and thought it would add a little extra dazzle.

You will need five matching lids, one which you will use same-size, and four others which you will have to cut down so that each successive one is ⅛" smaller than the previous one. I used 2¾" Ideal Dog Food lids for mine, and the pin came out a good size for the long scarves and great capes in vogue, but too large for the lapel of a suit or coat. However, if you were to turn over a little ¹⁄₁₆" hem, as it were, on the end of each petal, to make it smooth to the touch, then a 2¾" lid would be about

Top: **Dazzling Zinnia Pin.** *Bottom:* **Larger Cross Pendant.**

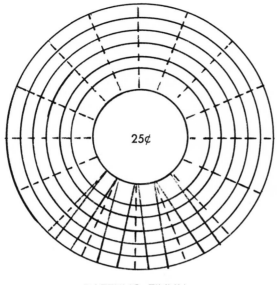

Divide lids into sixteen equal sections. (You will need five matching lids, one used as is; the others cut down so that each one is ⅛" smaller than the one preceding. The center of the smallest lid will be dime-size rather than quarter-size.)

Pare down sides of each section to make straight petals

DAZZLING ZINNIA

right. Plan to make earrings to match because this is too distinctive a design to go with other jewelry in your collection.

I did *not* hammer out the concentric circles in the lids because I wanted the ripply look, as did Tiffany. I just divided each lid into sixteen equal sections and pared down the sides of each section to make straight petals.

Note: As the layers become smaller, so will the center circle, so that the smallest layer will have a center the size of a dime, rather than a quarter.

When it came to assembling the layers, they somehow seemed more interesting to me in an assymetrical position, but that is a matter of personal taste, of course.

Holding all the layers in position, strike *two* awl holes through them and wire all together *through the pin finding* on the back, augmenting it with epoxy to make everything absolutely firm.

You will probably have to mail order the jewelry findings because the average dime store doesn't carry them. I would suggest you get the largest possible pinback—1½"—since it will be biting into thick material.

Isn't it smashing? Now all you need is the cape!

May Basket

This is such an appealing little thing that I keep it out on an end table all year round.

If you've made the Dazzling Zinnia, you'll know just how to pare down the sections of the lid to make straight-sided ribs, but in any case, the diagram shows you. The only difference is that you must have an *uneven* number of ribs in order to weave a proper basket.

I will tell you how to get an uneven number in a minute, but first select a lid at least 5″ across. A two-pound coffee-can lid will make a silver basket, or one of those luscious gold imported herring cans would make an enchanting gold basket.

May Basket

79

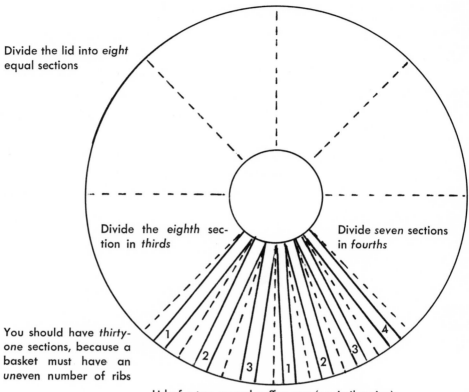

Divide the lid into *eight* equal sections

Divide the *eighth* section in *thirds*

Divide *seven* sections in *fourths*

You should have *thirty-one* sections, because a basket must have an *uneven* number of ribs

Lid of a two-pound coffee can (or similar size)
Pare down the sides of each section to make *straight* ribs

MAY BASKET

Hammer the lid perfectly smooth and divide it into *eight* equal sections. Then divide *seven* of those sections into *fourths*. *Divide the eighth section into thirds.* Can you see in the diagram that the ribs on the left are just a little fatter than the ribs on the right? That will give you thirty-one ribs and you'll be all set to weave.

Hook the end of your spool of Nu-Gold wire (or the end of your coil of No. 20 galvanized wire) around the base of a rib and squeeze it tight with the long-nosed pliers. (You may want to epoxy it in place.)

Weave under and over the ribs, pulling the wire tight enough to form a snug little nest. Bind the wire around the rim of the basket by bending the tip of each rib down over it.

For the handle, cut three 12″ lengths of wire; twist them together at one end with the pliers and tack them to your hammering board. Braid them together.

Nip off both ends of the braid neatly and feed them down through the openwork on opposite sides of the basket. Secure them with solder or glue, inside at the base of the ribs.

Note: You can make baskets from the *whole* can by dividing the sides of the can into an *un*equal number of sections (see chapter on Techniques) and weave them with ribbon or raffia, being careful not to let the edge of the ribs slice the ribbon as you slide it down into position. These baskets make charming containers for gifts from your kitchen and garden.

Bright Red Apple

The first Christmas tree was a balsam fir hung with red apples, the fir representing everlasting life, and the apples, the Garden of Eden and mortal sin. It was called a Paradise Tree and was the principal stage prop in medieval mystery plays.

Bright Red Apple

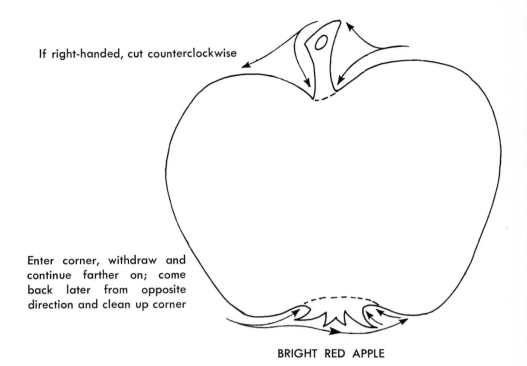

If right-handed, cut counterclockwise

Enter corner, withdraw and continue farther on; come back later from opposite direction and clean up corner

BRIGHT RED APPLE

This apple is my favorite ornament for the tree and I only wish you could see it in color. If there is the faintest ray of light in the room, these apples will reflect it and sing out "Merry Christmas!"

You will need lids 3½" in diameter, as well as bright red and Christmas green felt-tipped pens to paint them with. If you are going to make a lot of apples, you should use architect's linen for the pattern; it is much more durable than paper.

In any case, trace and cut out the pattern and apply it with rubber cement to the lid. Work around the pattern *counter*clockwise, bypassing the tight corners as recommended under Cutting Out Patterned Designs in the chapter on Techniques.

After peeling off the pattern and the cement, hammer around the edges and improve the outline by trimming off any roughnesses and cleaning up the corners.

Beat the apple from the back with the round end of the ball-peen hammer to give it texture and dimension, starting along the outer edge of the apple and working with many little hammer strokes gradually toward the center.

I used to beat the apples in my kitchen chopping bowl until I went right through the bottom of it. Then I discovered I could do almost as well on a flat board.

Leave the stem and blossom end *un*beaten, except to tap them from the *front* with the flat side of the hammer to make them seem to dimple-in realistically.

Paint your apples lustrous red, with a green stem and blossom end. Don't they look as tempting as Eve's?

8

WHAT ABOUT THE SIDES?

Now that we've taken a good look at a lid, let's consider the sides of cans. Except for the fact that some sides have ribs in them, they are—when opened up and flattened out—our basic source of plain, raw material. We are no longer circumscribed by a circle, much as we may appreciate the head start it gives us. We are now free to use this amorphous rectangle of metal in any way we choose. We can cut out all sorts of shapes and mold them to our purposes. We will find that most *useful* things are made from the sides of cans, like curtains, sconces, lanterns, and letterholders, but that lots of lovely, ornamental things are too. The wind chimes, especially, I think you will like, because they enchant the ear as well as the eye. So, reach for the raw material. You'll find everything you need to know about tackling it under Techniques.

Spider Sphere—Folksinger Holly Tashian's First Tin Construction. Ms. Tashian has put one of her guitar strings to unorthodox but equally lyrical purpose by knotting it to suspend a spider within spheres.

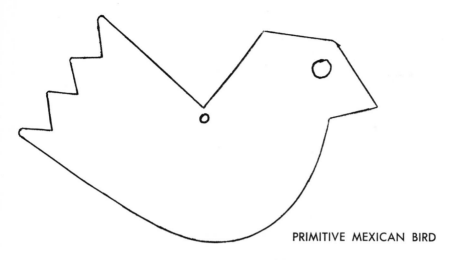

PRIMITIVE MEXICAN BIRD

Primitive Mexican Bird

You see these little birds used as decorative hooks for spoons and pots in Mexican kitchens. They are painted black and soldered onto an iron hook impaled in the white-plastered walls. They're so pert and cheery, I thought you might think of your own uses for them, and of course, they take only a minute or two to make. Wouldn't they be nice as gift tags at Christmas? And then your friends could hang them on their trees after they'd opened your present!

Just trace, cut out, make an awl hole, and color. I used Campbell's Soup cans which have squares and diamonds printed on the outside under the label. Then I painted within the squares and diamonds with Magic Markers in four different colors, like a patchwork quilt.

85

Bird in a Bush

Using a 46-ounce Campbell's Tomato Juice can, which is also marked in diamonds, you can make a mini-wall sculpture or sconce, using the seam as the trunk of the tree and gluing the bird and bucket to the top and bottom. Or a partridge in a pear tree perhaps?

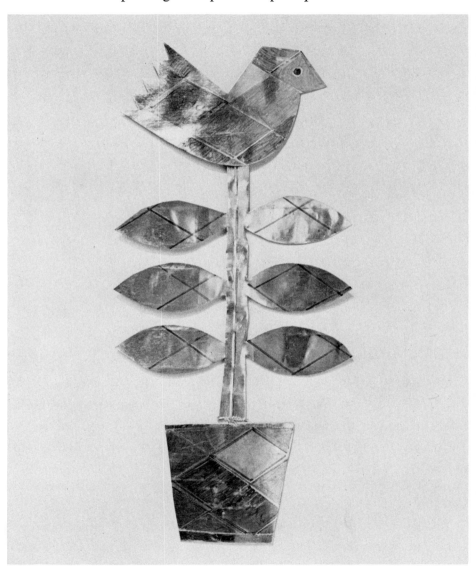

Bird in a Bush

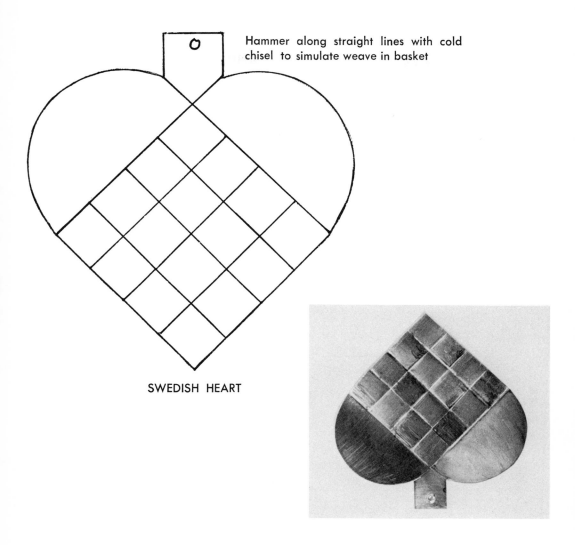

Hammer along straight lines with cold chisel to simulate weave in basket

SWEDISH HEART

Swedish Heart

Here again, I've used the Campbell's Soup cans marked in squares on the outside under the label to make a metal version of the Swedish, woven-paper, heart baskets. But, now that I think of it, you could make the baskets almost as easily without the special cans if you just have the cold chisel.

However, if you do find the special can, lay the pattern on it so the squares fit the base of the heart, as shown in the diagram. Cut the heart out and hammer along the lines with a cold chisel. Divide those squares in halves, with hammer and chisel, to simulate the weave in the basket. Paint the heart with Magic Markers in two contrasting colors—pink, of course, and green?

Bright Bird

Here's another cheerful fellow, a little more elaborate, because his wings are separate and must be glued on and then curved up, but all the same, easy to make.

If making lots of little birds, trace the pattern onto architect's linen (see Sources of Supply). Apply with rubber cement to the side of an *unbeaded* can. Cut out and hammer flat.

Fold the wings together, creasing well. Apply Weldit Cement or epoxy to the back of the bird where the wings will be attached, and to the inner side of the fold in the wings. Allow the glue to become tacky and slip the wings into position.

When dry, curve the wings up gracefully, rubbing them between your thumb and fingers to avoid the "bends." Test for the balance point (see

Bright Bird

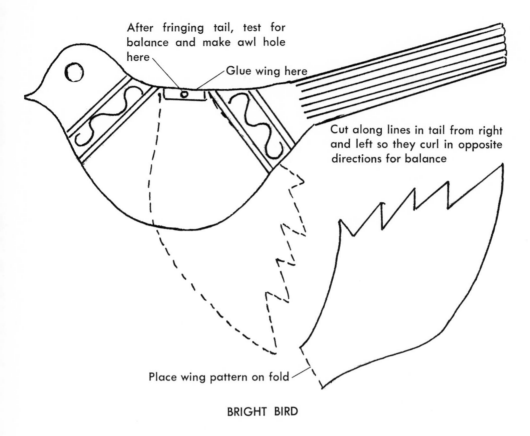

After fringing tail, test for balance and make awl hole here

Glue wing here

Cut along lines in tail from right and left so they curl in opposite directions for balance

Place wing pattern on fold

BRIGHT BIRD

Finding the Balance Point in Mobiles and Ornaments) which, for some reason, is very delicate in this bird. You may have to hang him from two holes. Fold the wings down again to strike the awl hole for hanging.

Fringe the tail by cutting narrow strips on the right side to the halfway point; then cut similar strips from the left side. Encourage the fringe to curl up and fan around over his back prettily. Make his eye with a paper punch if you have one, otherwise the awl.

Catbird

My live catbird might not recognize this portrait of himself with that disproportioned truncated tail, but all the same, a congregation of these tin catbirds cheers me up through the frosty winter, keeping me company

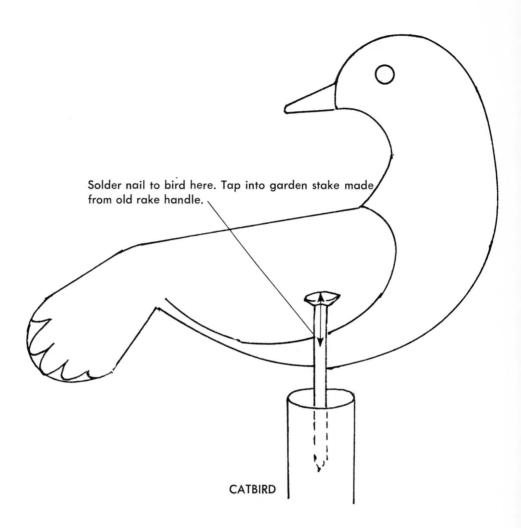

Solder nail to bird here. Tap into garden stake made from old rake handle.

CATBIRD

as they perch brightly on their stakes, watching over the beds covered with salt hay.

Apply the pattern with rubber cement to the side of an *unbeaded* can. In cutting the pattern out, make a clean curve around the tail, bypassing the individual feathers. You can delineate them later if you want to, but if they prove to be a problem, don't bother. The design is perfectly good without them.

Solder a sharp *finishing* nail (with a head) on the back (see Soldering). Paint him matte gray and black, or spray him with Krylon Crystal Clear, and tap him in place on the stake.

Gingerbread Man or Rag Doll

What kind of folksy Christmas tree would it be that didn't have dolls?

Cut these little people from the sides of *unbeaded* cans—unless you can find beaded cans that will stripe their shirts and trousers without also striping their heads and hands. We *can't* have them noodle-headed! They will seem more warm-blooded if they are gold-side front, so Campbell's Beef Bouillon is a good choice, or Green Giant Cream-Style Golden Corn.

If you plan to make lots of little dolls, use durable architect's linen for the pattern (see Sources of Supply). Place the fold of linen on the center line of the pattern in the book; trace the pattern onto the linen; then *glue the fold together* with rubber cement before cutting it out. That way it won't slip as you cut, and you'll get a perfect mirror image.

Gingerbread Man

91

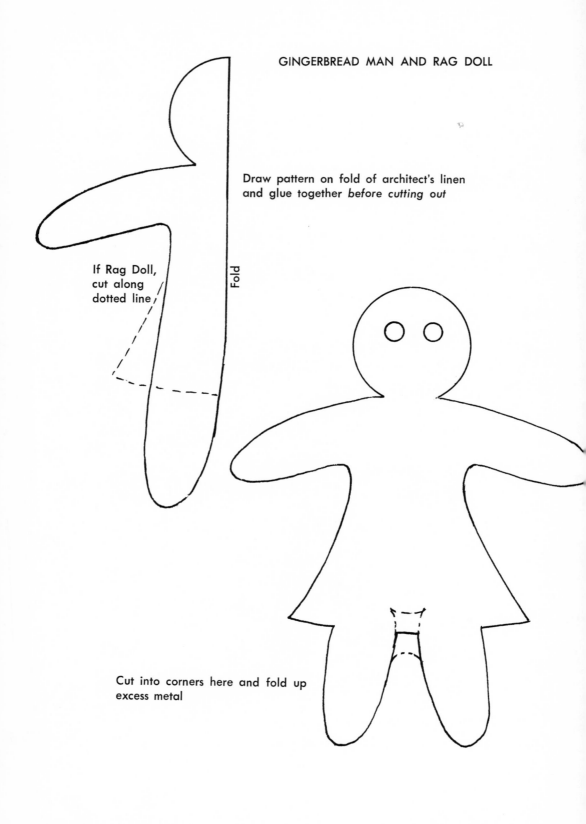

GINGERBREAD MAN AND RAG DOLL

Draw pattern on fold of architect's linen
and glue together *before cutting out*

If Rag Doll,
cut along
dotted line

Fold

Cut into corners here and fold up
excess metal

Now, open the pattern up and follow the usual procedure of applying it to the can with rubber cement.

If you have the time, it's fun to "dress" your dolls with Magic Markers, but it's the silhouette that really matters, so paint them as much or as little as you wish. The Gingerbread Man really needs nothing more than frosting-white eyes and buttons, although you can make paper-punch holes like those in the photograph instead. I tried adding braids and bows of red and green sock wool to the Rag Doll, but didn't like the wool on metal. Wire braids would be more in keeping, but that's fussy work unless, of course, you're making a gift and want it to be a treasure. Which, sometime, you might!

Diamond Belt

Just try this! It's so easy—an instant gift!

You may already know that many of Campbell's Soups come in cans with diamonds faintly, but precisely, printed on the outside *under the*

Diamond Belt and Earrings

Diamond Collar and Earrings

label. This means that to make a belt, you have nothing more to do than cut along the lines, round the corners of each diamond a bit, put awl holes in either end, and link them together!

You will need the sides of two such cans and approximately twenty-four oval jump rings (see Making Links), depending on your girth and the position in which you wear the belt—around the waist or the hips. I rather like the little Penobscot Indian Bell dangling on the end, but the belt looks quite finished enough without it; just remember to omit the *second* awl hole in that *last* diamond.

You may want to make the collar when you see how quickly the belt goes. You could do it easily in an afternoon.

Diamond Collar and Earrings

This is the *greatest!*

Before you begin, let me point out that with any design which has a lot of similar parts, the most efficient procedure to follow is to do as much of one type of operation at a time as is possible. Practice makes you good at it, and it goes along faster than if you interrupt your rhythm and start another type of operation. So, I suggest that you make all your

Borah tribeswomen wear diamonds even in their hair!

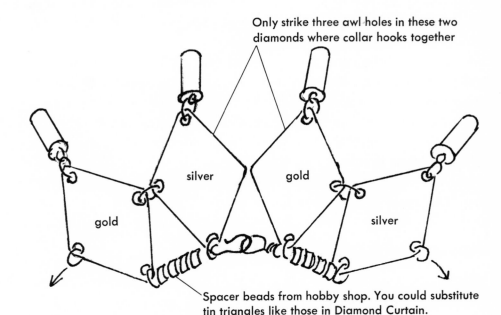

Only strike three awl holes in these two
diamonds where collar hooks together

silver   gold

gold   silver

Spacer beads from hobby shop. You could substitute
tin triangles like those in Diamond Curtain.

DIAMOND COLLAR

diamonds at once, strike all your awl holes at once, and cut all your
jump rings at once.

Also, in assembling the necklace, do it by units of diamonds and
cartridges, attaching all the cartridges to the diamonds before stringing
the units on the circlet. Then do all the *side*-linking at once, and last of
all, form the hook.

For both the collar and the earrings you will need twenty-seven gold
and silver diamonds. In eighteen of them, make four awl holes for linking,
striking nine from the gold side and nine from the silver side (see
Striking Awl Holes). Nip off the corners.

In the first and last diamonds on the circlet, make only three awl holes,
because, inasmuch as the necklace is to be clasped around the neck,
these diamonds will not be linked to each other and therefore do not
require the fourth hole.

In the remaining six diamonds, make only two holes, top and bottom,
since the ones for the necklace hang very nicely without being linked

on the side and the other two are for the earrings.

When striking the awl holes, always keep in mind which side will be *up*, the gold or the silver, because as you have probably discovered, awl holes look finished from the front but not from the back.

Next, make the twenty-four oval jump rings and attach them to the top of each of the diamonds.

Then, prepare the cartridges, perforating the center of the caps with a single awl hole.

Cut twenty-two lengths of wire, ⅝" long. Make a tiny loop with the round-nosed pliers on one end and pass the wire up through the hole in the cartridge. Form a *fat* loop above the cartridge; hook it into the hole in the bottom of the diamond; and squeeze it shut. Continue this process until you have all the diamond-cartridge units assembled.

For the circlet, cut a length of No. 18 Nu-Gold (or No. 18 galvanized wire) 16" long. With the round-nosed pliers, form a loop on one end. String the diamond-cartridge units onto the circlet with enough spacer beads in between to keep them sufficiently separated. If you don't like— or can't find—beads, you can substitute tin triangles, as in the Diamond Curtain.

When all the units are strung on the circlet, hold them snugly together by looping the free end of the wire *through* the last jump ring. Then form a hook as shown.

Link the diamonds together on the sides, and your collar is finished. The earrings are made the same way and attached directly to the little loop on the findings (see Sources of Supply).

Diamond Curtain

Because our house is darkened by evergreens which belong to our neighbors, we have not used conventional shades or curtains, but have softened the light at the top of the windows with a shallow heading of linked diamonds, a portion of which you see here. They would make a fabulous floor-length curtain, and I figure you'd only have to drink 128 cans of soup to be in business!

Diamond Curtain

Mobile-mask Pendant

This pendant is not only distinctive, but comfortable to wear. The choker you will have to buy. There is no way you can make that from a can, but my thought was that you probably already had one. The silver orbs are circles the size of a quarter, made from the side of the can, and the head and nose of the mask are made from the lid. Try to match the gold of the can to the gold of your choker (unless, of course, your choker is silver, which automatically eliminates the problem).

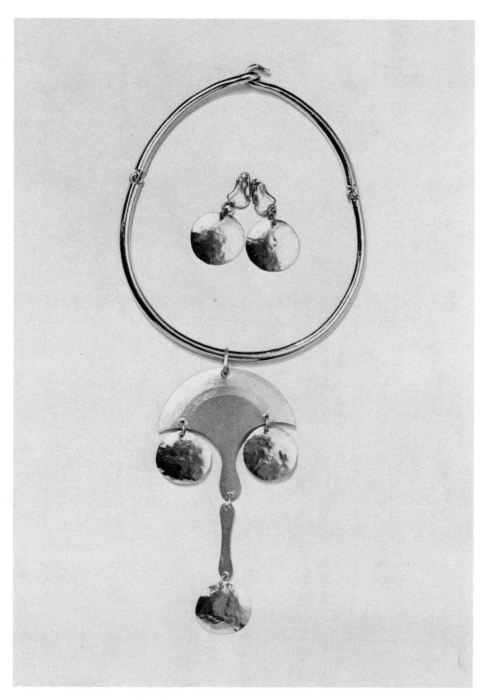

Mobile-mask Pendant and Earrings

You will also have to mail order earring findings unless your ears are pierced and you have wires available.

Superior snips are a must for this pendant; any *small, curved* piece requires them.

Hammer the lid on the silver side. Place it, gold-side up, on a piece of typing paper and trace around it. Lay the traced circle over the pattern in the book and draw in the design. Trace off the pendant pattern at the same time.

Cut out the circle and pendant patterns and apply with rubber cement to the lid and the flattened side of the can.

Cut the designs out *very* carefully. Remove paper and cement and hammer both from the *silver* side. Improve the outlines until you have achieved perfect symmetry in both pieces.

For the orbs, place a quarter on the side of the can and scribe around it with awl, or fine felt-tipped pen, five times. Cut the discs out and hammer them on the gold side with the *round* end of the ball-peen hammer.

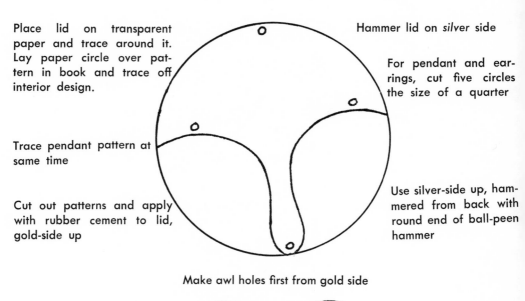

Place lid on transparent paper and trace around it. Lay paper circle over pattern in book and trace off interior design.

Trace pendant pattern at same time

Cut out patterns and apply with rubber cement to lid, gold-side up

Hammer lid on *silver* side

For pendant and earrings, cut five circles the size of a quarter

Use silver-side up, hammered from back with round end of ball-peen hammer

Make awl holes first from gold side

MOBILE-MASK PENDANT

Leaping Fish

Strike the necessary awl holes in all pieces and connect them with oval jump rings (see Making Links).

Leaping Fish

The fish and the sacred monogram, Chi Rho, are the most ancient symbols of the Christian faith and they make meaningful and graceful ornaments for the Christmas tree.

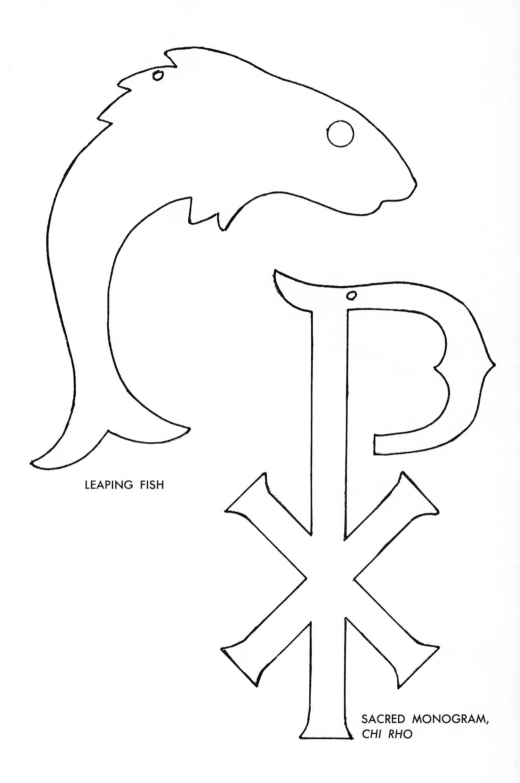

LEAPING FISH

SACRED MONOGRAM,
CHI RHO

Applying the fish pattern to the side of an *unhammered, beaded* can adds textural interest, don't you think? And the rhinestone glint in the eye does too.

Sacred Monogram, Chi Rho

The pattern for this ancient Greek symbol must be cut out with care to preserve the delicate contours. And you should have a pair of those superior Klenk snips to negotiate the curves. Use the sides of a Campbell's Soup can for this, and check out the balance point by the Scotch tape method.

Sacred Monogram, *Chi Rho*

Star of David

Star of David

When you consider the mathematical purity of the Star of David, you may be reminded that Edna St. Vincent Millay once remarked that "Euclid alone looked upon beauty bare." Certainly the simple strength of these eloquent equilateral triangles make them a fitting symbol for the greatest of Hebrew kings.

A one-gallon can will provide you with enough raw material for the star. Construct your triangle pattern by ruling off a line, 7″ long, near the bottom of a piece of typing paper. With your compass opened to fit the line, inscribe two arcs from either end of the line. Rule off the other two sides of the triangle; cut the triangle out; and apply it with rubber cement to the flattened side of the can. Cut two triangles. (See Star Sconce diagram if in doubt.)

With felt pen, rule margins on each of the triangles, ⅝" deep. Make a hole in the center of the triangles—large enough to admit your snips— by centering them over a crack and hammering the awl through the metal up to the hilt. Enlarge the hole with the round-nosed pliers, if necessary. Cut out the center of each triangle and glue the two together, with epoxy, in the form of a six-sided star.

Add a candlescroll to make a sconce, or superimpose it on the Star Sconce to add a subtle dimension (see Making Scrolls and Candlecups for special forked scroll).

Dingdong Bell Chime

The sound of this chime will remind you of cowbells—cool, distant, peaceful.

Start with the tier from which the bells will hang, by cutting down a 46-ounce V-8 Juice can to within ¾" of the bottom rim (see Cutting Down Cans), having already removed the bottom lid. Save the rest of the can for the bells and the fish.

Cut a strip of paper a little less than ¾" wide to butt around the sides of the cut-down can. Fold it in half three times and draw on it the curve indicated in the diagram.

While the paper is still folded, cut out that curve with scissors. Open the pattern up and apply it with rubber cement to the *inside* of the cut-down can so that one of the curves is centered over the seam. Cut along the curved line with snips. Strike an awl hole, from the inside, in each of the curves as shown.

To suspend the tier, you will need three awl holes, equidistant from each other, under the rim at the top. Take the paper pattern you have just used and fold it over on itself *in thirds*. Open it up; lay it on the tier, and mark the three points where the creases come. Strike the awl holes from the inside.

Dingdong Bell Chime

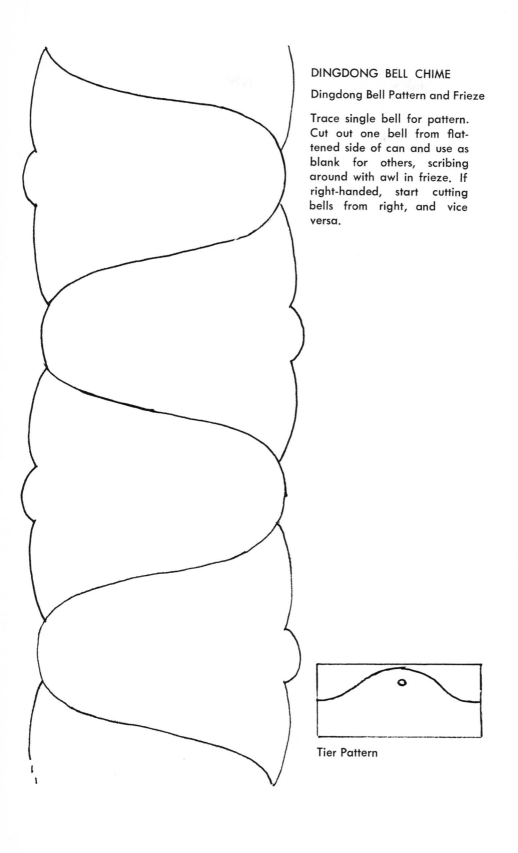

DINGDONG BELL CHIME

Dingdong Bell Pattern and Frieze

Trace single bell for pattern. Cut out one bell from flattened side of can and use as blank for others, scribing around with awl in frieze. If right-handed, start cutting bells from right, and vice versa.

Tier Pattern

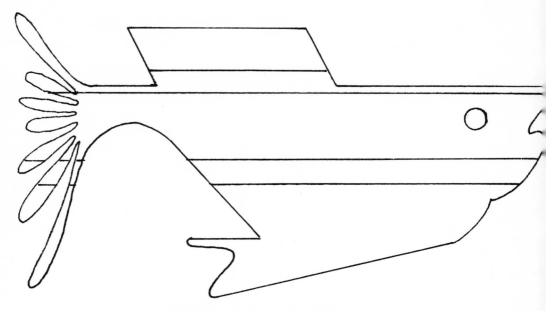

Oriental Fancy Fish cut from side of ribbed can

Cut three strands of tinsel cord 9″ long and tie them through the awl holes. Gather their tips together and bind them with fine hair wire 1½″ down, forming a loop for hanging at the same time. Suspend the frame from a hook so it will be handy to work on.

Next, hammer the sides of the can until nicely flat. Trace the bell and fish patterns onto a piece of paper. Cut them out and apply them with rubber cement to the side of the can, keeping in mind that you want to accommodate a frieze of eight bells in a row. This frieze will reduce your cutting by half and enable you to get the whole chime out of one can.

Cut out one bell and use it as a pattern to scribe around with the awl, forming a frieze as suggested by the diagram. If you are right-handed, start cutting the bells from the right end of the frieze. You may want to refer to Cutting Out Patterned Designs for help in negotiating the clappers.

Hammer your bells flat; improve the outlines; strike the awl holes for hanging.

Suspend the bells by metallic thread (*not* tinsel cord) on two levels, alternately 2″ and 3″ from the bottom of the tier. Trim the knots and daub them with epoxy.

Cut out the fish. Hammer him well. Make the eye with paper punch or awl. Read the instructions for Finding the Balance Point in Mobiles and Hanging Ornaments and suspend him with a 15″ length of metallic thread 9″ below the rim of the tier, wiring the tip of the strand to the others at the neck.

Cut two 12″ lengths of tinsel cord for bows and tie them in place on either side of the neck with another 6″ length of cord. Make three similar but smaller bows and wire them to the tier at the three points where it is suspended.

If the chime is to swing on an open porch, protect it with Krylon Crystal Clear and it will carol-ling-along for years.

Cruising Fish Wind Chime

These fishy friends clang sonorously as they cruise around under the overhang of Elinor Ehrman's lakeside summer house.

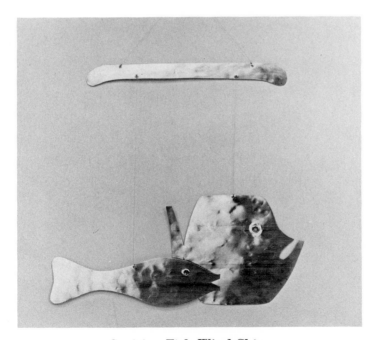

Cruising Fish Wind Chime

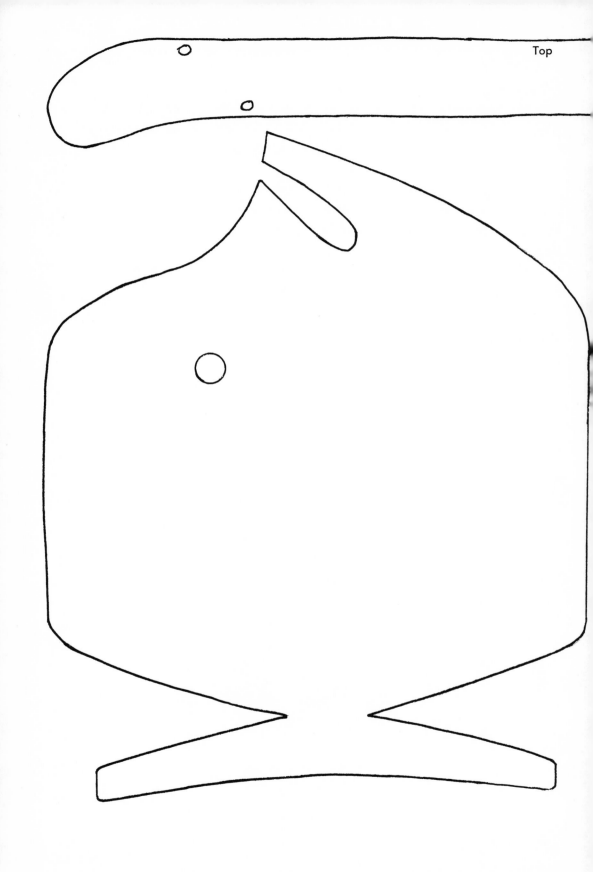

Top

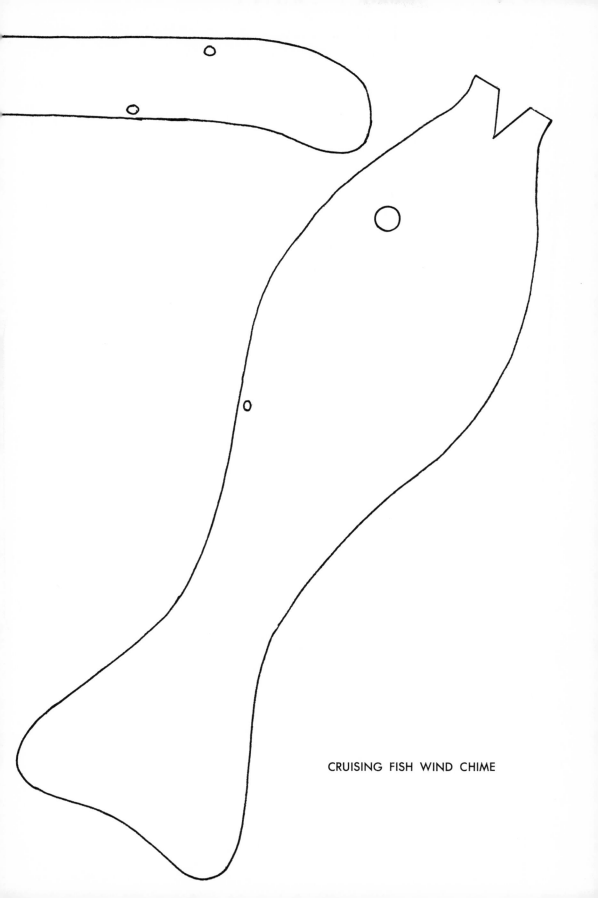

CRUISING FISH WIND CHIME

To make the chime, I used two sides of a five-gallon soybean oil can, but you could also make it from any one-gallon can by shortening the crossbar a little, or by using a driftwood branch to float them from instead.

There is no how-to called for in this piece to speak of. Just trace the patterns and cut the fishes out. Hammer them well all over, with the flat end of the ball peen.

To make suitably large eyes, strike the awl over a crack between kitchen counters or the hole in your spool of Nu-Gold wire, forcing it down to the hilt. If that hole isn't big enough, insert your round-nosed pliers in the eye and rotate them to enlarge it.

Because the square fish is heavier than the slim one, you may have to *off-center* the loop for hanging just a little bit. No one will ever notice.

Dove Chime

These fluid forms were originally designed in stained glass, but they are very effective in metal and the slightest tremor, like a car passing, will set them singing.

Dove Chime

112

DOVE CHIME

DOVE CHIME

As with the Cruising Fish, there is no special instruction needed. Just cut out the forms from any *unbeaded, unlithographed* can large enough to accommodate the dove pattern. Hammer and polish them well.

Strike the awl holes and suspend them on fishline from a brass tube or dowel. If they are going to be hung outdoors, they should be sprayed with Krylon Crystal Clear, of course. Both in form and sound, you will find them perpetually pleasing.

Japanese Garden Lantern

Japanese Garden Lanterns

Even if you have spotlights on the trees, you still need candles on the table when you eat out on a summer's evening. These lanterns will give you all the illumination necessary and cast pretty shadows as well.

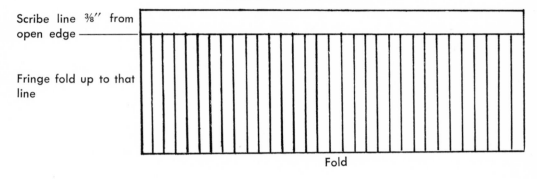

The side of one can of cherry pie filling folded in half and trimmed "true"

Scribe line ⅜" from open edge

Fringe fold up to that line

Fold

GARDEN LANTERN

When I am having a number of guests, I often make two large cherry or strawberry tarts, using *canned pie mix,* and this means that I have *four* cans, right there, to make lanterns with. And they only take a jiffy!

Remove the top rim and the whole bottom end of the can, saving the end for the base of the lantern.

Trim off the seam, hammer the sides flat, and fold the piece in half lengthwise, creasing it well with the hammer. Trim the piece "true," using a shirt cardboard as a handy right-angle.

Along the *open* edge, scribe a margin with the awl, ⅜" deep. *Fringe the fold* by making straight cuts every ¼" up *to* the margin. Hammer the fringe flat, working back and forth from the base of the cuts to the fold line.

Force the fold open with a broad cold chisel or kitchen spatula. Pull the sides apart with your hands as far as possible.

Turn the fringed piece over, *fold up,* and grasp one end with both hands. Press the metal between your thumbs and fingers to make it curl *down under.* Continue the full length of the piece till it is round like a lantern. Glue the overlap with epoxy, clamping it together until dry. (If in a hurry, wire it together through awl holes top and bottom.)

Seat the lantern in epoxy on the end of the can, *rim up,* and spray-paint it with Krylon Flat White, inside and out. Set a votive candle inside for a flattering flicker of light.

Desk Pad and Letterholder

The essential beauty of heavy tinplate is instantly apparent in these desk accessories and their purity is such that there seems little need for ornamentation.

For the pad itself, cut a piece of Masonite 12″ x 19″, or whatever special size you need.

For the corners, cut four small oblongs, 5″ x 2½″, from the sides of a five-gallon can, burnished with 0000 steel wool.

Place each oblong, in turn, across the corner of your hammering board so that the middle of one long side is centered on the corner, forming

Desk Pad and Letterholder

Center piece on corner of hammering board and gently hammer flanges down at right angles as indicated by dotted lines. Turn over, flanges up. Slip Masonite into position and press flanges down onto it.

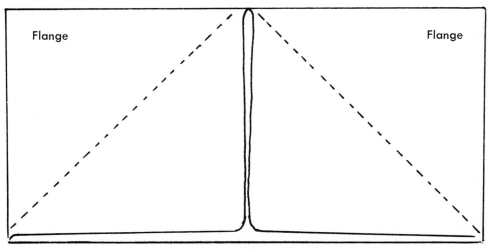

Round corners and trim edges smooth and even

DESK PAD

equal overhanging triangles. Carefully hammer down these two flanges against the sides of the board.

Turn the corners over, flanges up; slip the Masonite pad into position, and press the flanges over onto it. Remove the Masonite; tap the folded flanges gently where they need it to make them neat and true. Trim off the points on the corners and any superfluous metal so that all is smooth and symmetrical.

For the Letterholder, cut a large oblong, 12" x 7" from the same can. (Your desk set should match.) Use a shirt cardboard or other reliable right-angle to make sure the piece is true.

Apply the Letterholder pattern first to one end of the oblong and then the other with rubber cement, and cut out the curvilinear design.

To shape the *curved sections* of the Letterholder, first nail a short length of broom handle (or similarly-sized dowel) to one side of your hammering board. Then bend the metal over it in easy curves. You may need to hammer it *gently* to get the curves started. Notice that the front

118

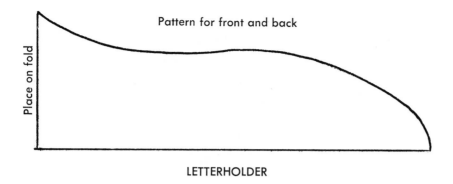

Place on fold

Pattern for front and back

LETTERHOLDER

of the holder is low; that the separating curve is higher, and that the back panel is highest.

Wouldn't these accessories be stunning on a white formica desk?

Colonial Drawer Pull and Tieback

Our unpainted kitchen cabinets came with wooden and metal handles too pedestrian for me, so I made copies of old ones out of the sides of cans. They are more than strong enough for the purpose because the handle section has been reinforced by turning over the edges, like a candlescroll.

Trace the pattern and cut it out carefully, applying it with rubber cement to the sides of any *unbeaded* can. Take special care in cutting the corners where the hearts join the handle. To look professional, those measurements must be true and identical, top and bottom. Fuss with them until they are alike.

Then, with your long-nosed pliers, bend down the two *corners* of the handle *on the right side at right angles* and lay it along the straight-edge of your hammering board. Hammer down the rest of the overhang. Turn the handle on its back and carefully hammer the overhang flat. Repeat on the other side. Bend the handle around a lolly column or large pipe until it is properly curved.

Strike awl holes in the hearts. Spray with flat black or tôle red.

The tieback is made the same way but with only one heart, from a strip 10″ long. Screw it to the wall through two awl holes in the plain end.

Center: **Crimped Sconce.** *Left:* **Colonial Tieback.** *Right:* **Colonial Drawer Pull.**

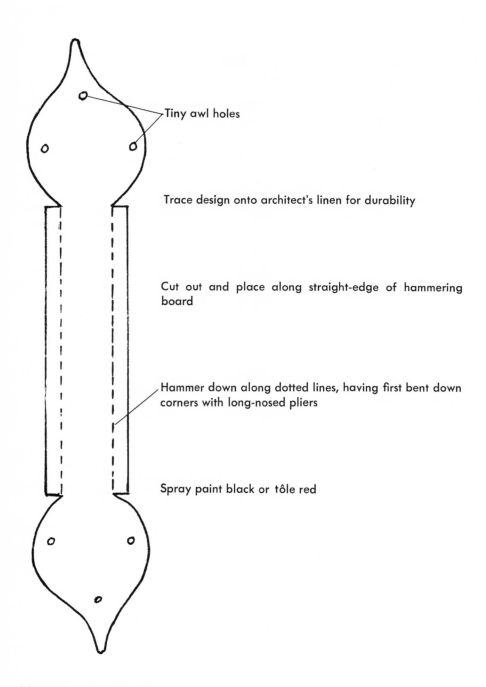

Tiny awl holes

Trace design onto architect's linen for durability

Cut out and place along straight-edge of hammering board

Hammer down along dotted lines, having first bent down corners with long-nosed pliers

Spray paint black or tôle red

COLONIAL DRAWER PULL

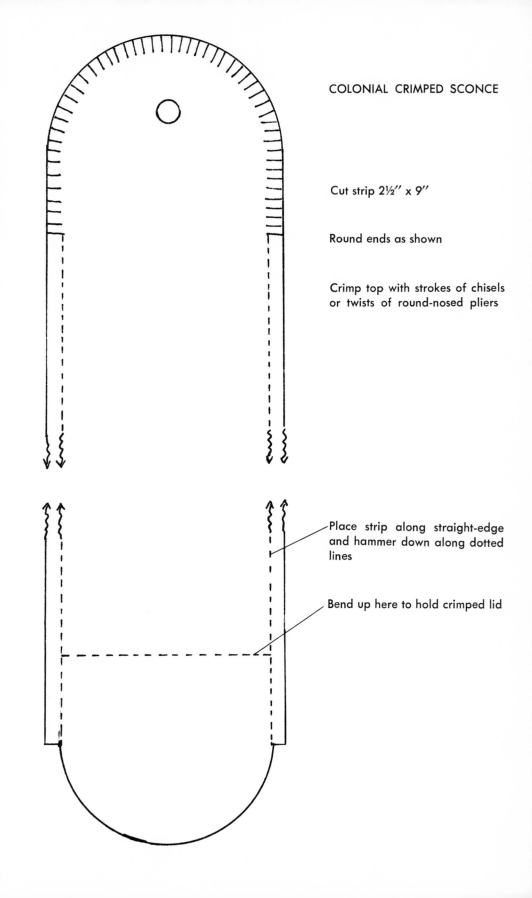

COLONIAL CRIMPED SCONCE

Cut strip 2½″ x 9″

Round ends as shown

Crimp top with strokes of chisels
or twists of round-nosed pliers

Place strip along straight-edge
and hammer down along dotted
lines

Bend up here to hold crimped lid

Crimped Sconce

If you pursue an interest in antiques and authenticity long enough, you will finally discover that our forebears used whatever they could lay their hands on. There was no *one* way to make something. Each maker used the materials and tools he had on hand, and no matter how crude his end-product was, it is now considered an antique and costs money!

In shows and shops, you'll come across sconces that are nothing more than a narrow strap with a candlecup on one end. Another will consist of three straps tacked side by side onto a board. Crimped sconces were made short and narrow, short and fat, or tall and narrow. Some have been turned and grooved by machine to make them rigid, while others were crudely struck with a chisel. I mention these differences only to encourage you to do your own thing. Let the space your sconce will hang in determine its proportions and let the style of your furniture determine the design.

The Crimped Sconce you see here is small, made from a piece of un-beaded tinplate 2½" x 9". The top was cut in a curve and crimped with round-nosed pliers (see Crimping). The sides were folded over like a candlescroll (see Making Scrolls and Candlecups). The whole sconce was sprayed with Illinois Bronze Tôle Red.

Two Easy, Easy Sconces

These sconces are practically ready to hang on the wall when you've cranked off the ends of cans with the opener. In fact, the only thing that really takes time is consuming the coffee and olive oil that comes in them, first!

To make the round sconce, remove the *whole end* of a three-pound coffee can (see Removing Rims). Beat the center section with the round end of the ball-peen hammer *from the back* on a flat board, carefully, so as not to spoil the perfection of the concentric circle.

Cut the pull chain to size and glue it in place with epoxy. Fashion the candlescroll and candlecup (see Making Scrolls and Candlecups); glue it on with epoxy, and your sconce is finished.

The crimped sconce is even easier and it wears well, blending nicely into the background.

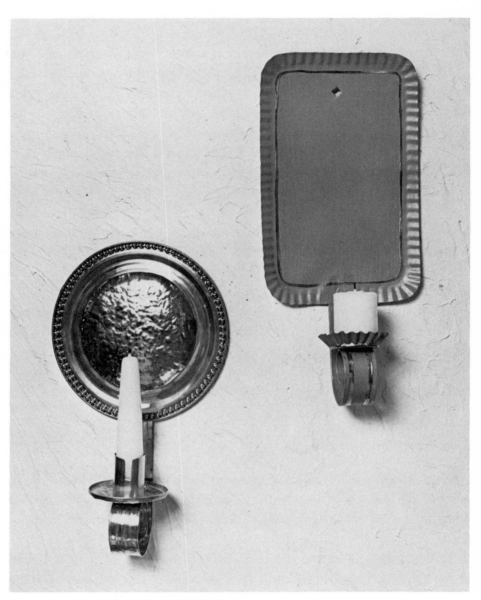

Two Easy, Easy Sconces

Remove the bottom lid of a gallon can and crimp the edges with round or long-nosed pliers (see Crimping). Glue on the candlescroll, which, you notice, is here supplied with a crimped cup to hold a votive candle. Add an awl hole for hanging.

This sconce is painted Early American Mustard, and ornamented with a gold borderline, but you could decoupage a flower on it, or glue a mirror there, if its simplicity seems too stark to you.

Star Sconce

You will enjoy making, and living with, this classic star sconce.

It doesn't really matter whether you use a lid or the sides of a can for this sconce, but whichever you choose, it must be large, like the lid of a commercial flounder fillet can or the sides of a one-gallon can.

Draw two 7″ equilateral triangles on a piece of typing paper as shown in the diagram, and cut them out. Glue them together with rubber cement to form a six-sided star. Using this as a pattern, apply it with rubber cement to the piece of tinplate and cut it out. Hammer the star well.

Star Sconce

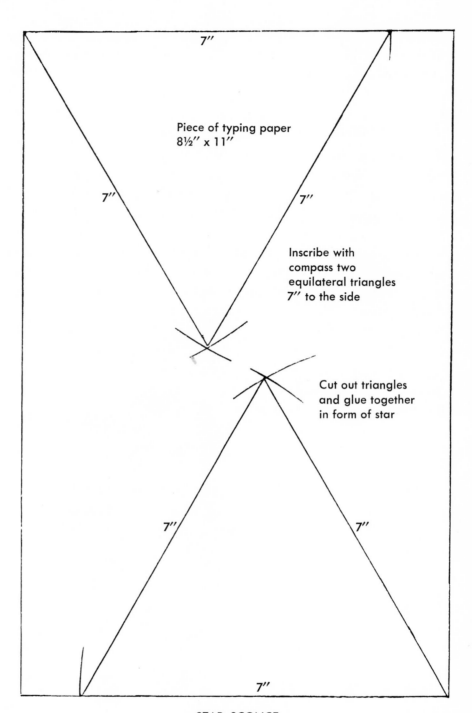

7"

7"   7"

Piece of typing paper
8½" x 11"

Inscribe with
compass two
equilateral triangles
7" to the side

Cut out triangles
and glue together
in form of star

7"   7"

7"

STAR SCONCE

Find the center of the star with the compass or by eye, and draw lines to each of the six points with felt pen. Score along those lines with hammer and cold chisel; or, if lacking the chisel, glue your ruler on the line with rubber cement and score heavily along it with the awl.

Very carefully, accentuate the center point with a nail set from the front; or ornament it with an upholsterer's tack. Craft a candlescroll with a loop for hanging. Hammer one point of the star flat and glue the candlescroll to it with epoxy.

Penobscot Indian Bell Wind Chime

Curiously enough, these narrow bells, which I first discovered at a Penobscot Indian festival in Maine, I have since found in other Indian cultures as far distant from each other as Arizona and Peru. Apparently the design is as old as the civilizations themselves, and probably the reason the Indians go to the trouble of making them so long and narrow is because they are more practical to wear than short fat ones would be.

Wear them the Indians do, in clusters of three on their deerskin dresses and jerkins, and you've never heard so gay a sound as all these little bells tinkling coolly together on a hot August afternoon.

This wind chime was the first thing my children and I ever made out of tin cans, which should cheer you on. In those days certain coffee cans came with detachable rims, all neatly stapled together, and naturally, I availed myself of them. But they have since vanished down the path of progress and you will have to make your own, either by cutting down a can, as in the Dingdong Bell Chime, or by cutting ribs from the sides of cans and forming hoops. I like to fold over the edges of the hoops, as you would a candlescroll, but of course it's more work and not really necessary if the strips are accurately cut and well-hammered. Suit yourself.

To make this wind chime, you will need the lid and the sides of a beaded, all-silver can, 6″ in diameter, such as Idahoan Instant Mashed Potatoes. You will also need two 5″ and two 4″ lids (probably from two-pound and one-pound coffee cans), as well as a hank of dark green rug wool from the dime store.

To make the hoops, cut three beaded strips ½″ wide (or ¾″ if folding

127

Penobscot Indian Bell Wind Chime

over the edges) from the sides of the large can. The first strip should be 16½" long, the second 12½", and the third 6½". Hammer them well and recurve them around another can.

To fasten the hoops together, strike two awl holes at one end of each strip, one above the other. Bring the other end of the strip around to overlap ½" and strike corresponding holes in it *through* the first set. Cut little U-lengths of wire and bind the hoops together. Hammer the joining flat, from the inside.

In order to suspend the hoops, strike three equidistant awl holes near the edge of each hoop. If your judgment is good, you can do this by eye, but if you want to make sure the holes are truly equidistant, cut a strip of paper to butt around the hoop, and fold it in thirds. Mark the points on the hoops where the creases come and strike awl holes at those points from *inside* the hoop, so as not to bend it out of shape.

Cut three 7" strands of yarn to suspend the largest hoop, three 8" strands for the middle hoop, and three 9" strands for the smallest hoop. Knot one end of each strand. Feed the strands, *from the inside*, through the awl holes in their respective hoops. Set aside.

Now cut the strands from which the *bells* will hang, because they must be slipped inside before the bells are closed tight.

Cut eighteen 9" strands of yarn and knot *both* ends. (Later you will attach bells to *each* end of these strands.) Cut a 15" strand for the low center bell and knot *one* end. Cut two 3" strands for the bells at the top of the chime and knot one end.

Now make the bells: divide all five lids into eight equal sections (see Dividing a Lid into Equal Sections). This will give you forty pie-shaped pieces. (Since you need only *seven large* bells, use the eighth piece to experiment with.)

Clip off the point on each of the pie-shaped pieces in a curve as shown, and hammer them with the flat end of the ball peen *until they curve up.*

Continue the curve by grasping the very outer edge of the bell with the *round-nosed pliers*, palm toward you. With little *inward twists of the wrist*, work across the bottom of the bell, *pinching and twisting your wrist*, moving along almost imperceptibly until you can go no further.

Turn the bell over and work the other way, again grasping the very

Tier

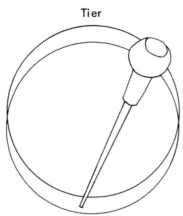

Bell

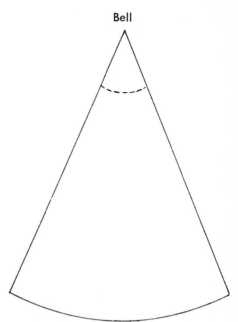

1. Cut ½″ wide strips from sides of beaded can; curve around smaller can; strike awl holes in ends and fasten with U-shaped wires

Strike three equidistant awl holes from inside to suspend hoop

2. Nip off top of bell
Hammer with flat of hammer on board till bell curves up

3.
Grasp right edge of bell with round-nosed pliers and continue inward curve with twist of wrist

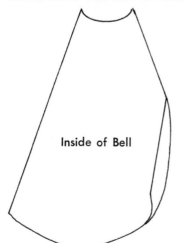

Inside of Bell

4.
Turn over and repeat in opposite direction with outward twists of wrist

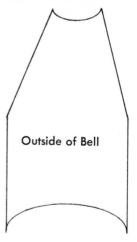

Outside of Bell

PENOBSCOT INDIAN BELL WIND CHIME

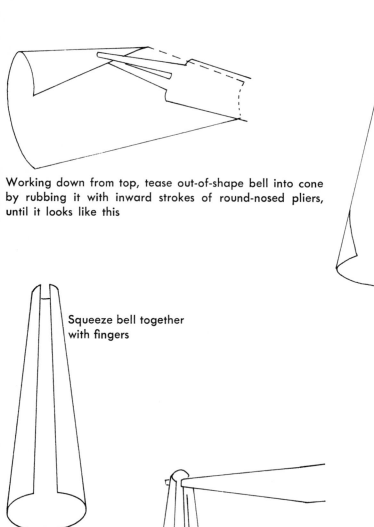

Working down from top, tease out-of-shape bell into cone by rubbing it with inward strokes of round-nosed pliers, until it looks like this

Squeeze bell together with fingers

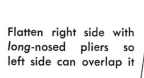

Flatten right side with *long*-nosed pliers so left side can overlap it

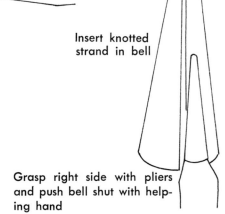

Insert knotted strand in bell

Grasp right side with pliers and push bell shut with helping hand

PENOBSCOT INDIAN BELL WIND CHIME

outer edge of the bell with the pliers and completing the curve this time with *outward* twists of the wrist.

Because the bell is too long for your pliers, you will have been able to curve only the bottom portion of it, and now you must pull the top into shape by rubbing the bend out of it with the pliers. With similar little twists of the wrist, first on one side and then on the other, *coax it into a cone.*

When you have finally established a perfect curve, you can almost squeeze the cone together with your thumbs and fingers, but one side must pass under the other in order to overlap properly. So switch to the long-nosed pliers and flatten the right-hand edge of the bell just enough to allow it to pass under the left edge.

Before closing the bell, insert the knotted strand. Then grasping the flattened side of the bell with the long-nosed pliers, push the left side of the bell over the right side with the fingers of your helping hand, thus making it overlap. Squeeze the overlap tight with the pliers.

Now look at your bell from the bottom to see if it's round. If you find a bend or crease anywhere, flatten it with the long-nosed pliers. If that opens the bell up again, squeeze it shut with your fingers. Fuss over it until you get it quite perfect. Don't be discouraged; the first bell is the hardest. Pretty soon your fingers will get the hang of it and *you'll* be watching television while *they* do the work!

Altogether you will make sixteen small bells, sixteen medium bells, and seven large bells. Don't forget to insert the knotted strands as you go along, *with matching bells on either end of all but the three special strands.* Attach one large bell to the 15″ strand and two large bells to the two 3″ bow strands. Fringe these bells ½″ deep.

Hang *eight* strands of *small* bells over the *largest* hoop, spacing them evenly and using a single knot to hold them in place.

Hang *six* strands of *medium* bells over the *second* hoop and knot them in place. Finally, hang *two* strands of *medium* bells and *two* strands of *large* bells on the *lowest* hoop, alternating the sizes, and knot them in place.

Now, put it all together! Pick up the 15″ strand with the low center bell and pass it up through the smallest tier. Grasp the four strands by

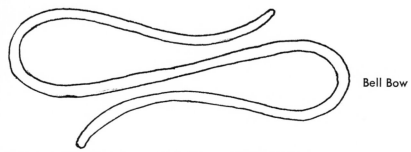

Bell Bow

Cut two 14″ strands of yarn for bows and fold them like so. Using a separate 6″ strand of yarn, catch them in place at neck of chime, along with fringed bells.

their tips and pass them up through the second tier, et cetera. When you get everything pulled up through, lay the chime down and gather all the tips together so they are more or less even and organized. Bind them 1½″ down from the top with fine wire, making a loop for hanging in the process.

Hang the chime from a handy hook to see if the tiers are level. Make adjustments, if necessary, by cautiously pulling *down* on the rims where they are too high *while holding the neck firmly.*

Trim the neck with bows and bells and let them all ring out!

Suggestion: Tiny bells made from soft gold Campbell Soup lids make charming bangles for earrings and bracelets. Here's the pattern showing you how to cut the bell for linking.

PENOBSCOT INDIAN BELL EARRINGS OR BRACELET CHARM

Divide Campbell's Soup can lid into eight pie-shaped pieces. Cut as shown.

Strike awl hole before shaping bell

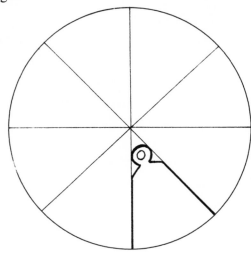

The Empire State Building on a Starry Night—another of Holly Tashian's first imaginations in tin.

9

SUPERSTRUCTURES
FROM THE WHOLE CAN

By the simple expedient of dividing the sides of cans into equal sections and pulling those sections out horizontally, like the rays of the sun, you add the height of the can to its width, thus doubling or trebling its dimensions. If you use one of those gorgeous gold 10″ x 12″ commercial salad cans from the A & P in this fashion, you come up with something pretty sensational, like the Celebration Chandelier on page 146.

You can also double the dazzle by placing one star on top of another, or on the tips of another. These Superstructures can be small and exquisite or large and spectacular. They may look complicated, but they are essentially simple. Just look to Techniques for tips in getting you started.

Sunflower Sconce

This shaggy sunflower is nothing more than two Bumble Bee Tuna cans back to back, with their sides divided into equal sections and fringed. But you do need *serrated* snips!

First, remove the rims from the tops of the two cans; then divide their sides into approximately 1″ sections (see Dividing the Sides of Cans into Equal Sections). Hammer these sections out to the side and fringe them, placing the fingers of your helping hand in the way of the strands as you cut, so they will *not* curl. Tap the fringe gently with the hammer to straighten it a bit, but do not flatten it altogether, for your sconce will look more like a sunflower if it's shaggy.

Sunflower Sconce

Glue the two fringed cans back to back with epoxy, so that the gold one lies behind the silver one. Simulate the seed head in the center by fringing one of the lids right up to the innermost concentric circle and beating it on the *silver* side in a wooden chopping bowl with the round end of the ball-peen hammer.

Place this domed lid, gold-side up, in the center of the sunflower and push it down until it *almost reaches the rim*. Remove it while you apply epoxy all around the inside of the rim, and pat the pull chain (cut to fit) in place. Apply epoxy to the tips of the fringed lid and replace it in the center of the sunflower.

While all this is drying, make the candlescroll (see Making Scrolls and Candlecups). You will probably not find a candlecup in the dime store like the one in the photograph, but you can make your own very easily.

Glue the scroll to the back of the sunflower, and form a loop for hanging by bending the tips of two pieces of fringe together with the long-nosed pliers.

Frosty Treetop Star

Add the sparkle of a star to the golden rays of the sun and crown the glory of your Christmas tree!

You will need a gold-lacquered Campbell's Soup can (Beef Bouillon is always good) and the lid from a two-pound coffee can. If you haven't already made a Frosty Star, don't be put off by all those curls. They are incredibly easy to make: you just cut straight ahead, and the little curls go spiralling off. Follow the directions on page 68.

To make the golden sun-rays, remove the bottom lid of the soup can and divide the sides into equal sections approximately ½" wide (see Dividing the Sides of Cans into Equal Sections). Save yourself unnecessary cutting by ruling off the rays in each creased section *before* you glue the paper to the can. Cut along the *ruled* lines instead of the creases

Center the Frosty Star over the sun-rays, and holding it all firmly in place with your helping hand, pull four of the curls through to the back near the base of the rays with long-nosed pliers, so the star is hooked to the rays. For added security, squeeze a drop of glue on each hook.

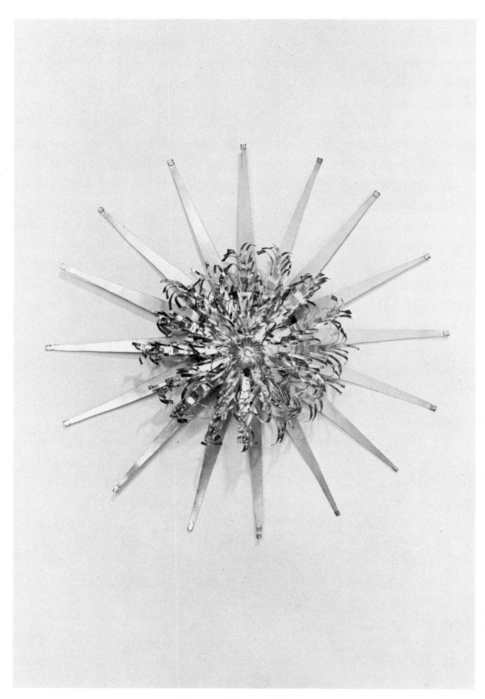

Frosty Treetop Star

Divide the sides of a can into equal ½" sections and rule off rays on the folded piece of paper before gluing to can. Cut along rays instead of creases.

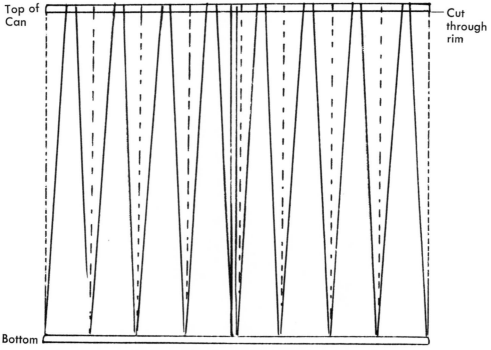

Top of
Can

Cut
through
rim

Bottom

Center one ray on seam for rigidity when mounting star on Christmas tree

FROSTY TREETOP STAR

Dip the tips of the sun-rays in silver glitter and wire your star to the treetop with a shining light behind it.

Sizzling Sun Sculpture

You can *top* stars with stars or *tip* them with stars. This conception of the sun has been a favorite of people everywhere in the Western World for centuries. How do you suppose the ancients ever discovered that some of the rays emanating from the sun were cold and others were hot? Anyway, that's what those straight and wavy lines signify. Imagine what a handsome mirror this would make if you used that huge fillet of sole can! When you divide the sides of the can into equal sections, draw the

139

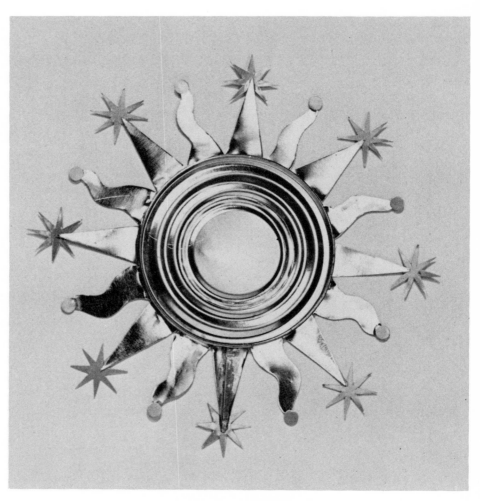

Sizzling Sun Sculpture

straight and wavy lines on the folded paper before you glue it on the can, and cut along them instead of the creases (see the chapter on Techniques, if necessary). If your can is one of the gorgeous gold ones from the delicatessen department in the A & P, make silver stars for the tips to provide contrast. I'm sure I don't have to remind you to keep all the elements to scale.

Speaking of scale, the Mexicans make really mini-mirrors in the form of sizzling suns, not more than 4″ overall. A friend of mine uses hers as a pin tray on her bureau.

Mexican Sun Mirror

Another dazzler! It will brighten the darkest corner. Scallops came in this beauty that resembles a gold paint pail minus the handle. You will need two to make this effulgent sun.

Divide the sides of *one* can into equal sections (see Dividing the Sides of Cans into Equal Sections), ruling off the rays on the folds of the paper before gluing it to the can with rubber cement. Cut along the ruled lines; remove the paper and cement, and hammer the rays out to the side to form a halo.

Mexican Sun Mirror

141

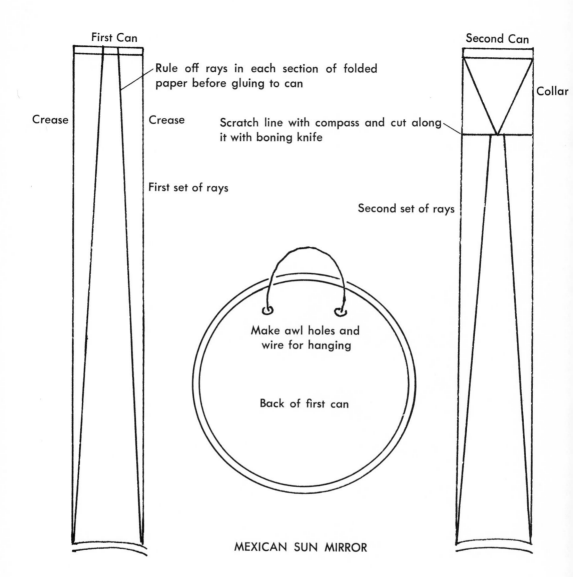

First Can

Rule off rays in each section of folded paper before gluing to can

Crease

Crease

First set of rays

Second Can

Collar

Scratch line with compass and cut along it with boning knife

Second set of rays

Make awl holes and wire for hanging

Back of first can

MEXICAN SUN MIRROR

With a compass, mark off a line around the side of the *second* can, 1½″ from one end. Plunge a boning knife into the side of the can on that line and cut around the can along the line, as you would a loaf of bread.

Divide the sides of both these pieces into equal sections to form a second set of rays and a collar, as shown in the diagram. Wire awl holes for hanging in the back of the first can and glue all the layers together with epoxy.

You might want to glue a pull chain in the groove around the mirror. Which reminds me, your mirror man will be grateful if you make him a paper pattern; and, for your own sake, beseech him to give you some mastic for gluing the mirror in place. Other glues lift the silver.

Frosty Evening Star Sculpture

This sculpture deserves to be displayed on a wall where it can shine in solitary splendor, as it does in the home of Mrs. John E. Massengale III

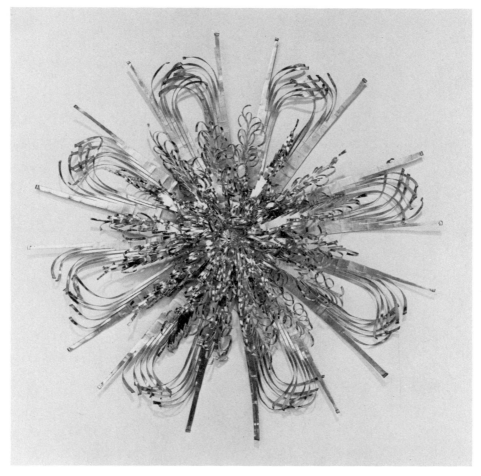

Frosty Evening Star Sculpture

143

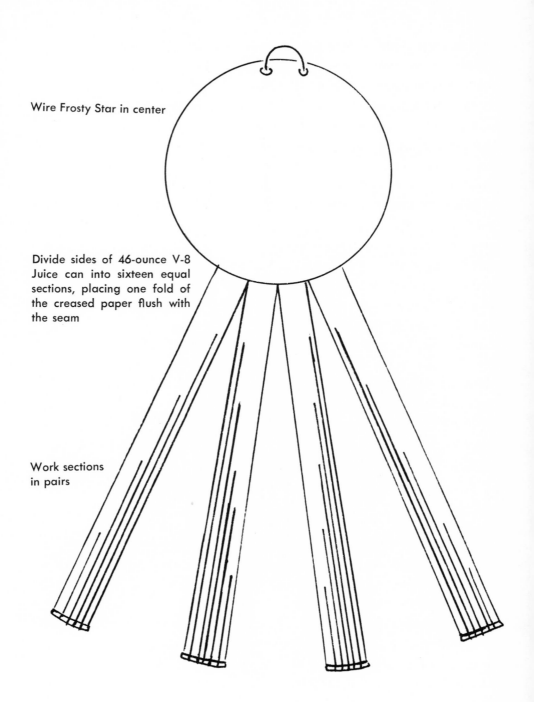

Wire Frosty Star in center

Divide sides of 46-ounce V-8
Juice can into sixteen equal
sections, placing one fold of
the creased paper flush with
the seam

Work sections
in pairs

FROSTY EVENING STAR SCULPTURE

over the grand piano, in company with tapestries, crushed velvet, and crystal.

The fabulous Frosty Star in the center is made from the famous Fulton Fish Market flounder can. The Evening Star on which the Frosty is set is made from the whole 46-ounce V-8 Juice can, measuring 18″ in diameter when opened up in this fashion. You will find directions for the Frosty Star on page 68, but I will give you instructions for the Evening Star here.

Divide the sides of the large V-8 Juice can into sixteen equal sections (see Dividing the Sides of Cans into Equal Sections). Hammer them out to the side and burnish them with 0000 steel wool till they shine like satin.

Cut each section as indicated in the diagram, *inhibiting the curls by holding fingers of your helping hand in the way.* To flatten the strands along the shaft, run your hammer up into them until they are *almost* back in place. Hammer the shafts from both sides.

Spiral the tips of the strands over your fingers in the direction they seem to want to go, encouraging the paired sections to face and almost intertwine with each other. Fuss with them affectionately until their curves are identical.

Secure the Frosty Star to the Evening Star with epoxy at a number of points where they touch. Protect the sculpture forever with Krylon Crystal Clear.

Sole Star Sculpture

Here is that fantastic fillet of sole can again. The bottom lid was used for the Frosty Star in the center. The sides of the can have been divided into equal sections too large to be bent out without buckling *unless the rim was flattened.*

To accomplish this, you must place the can (whose sides you have already divided and cut to a point, but *not* opened out) *far enough over the corner of a kitchen counter* to accommodate the width of *one* section. Hammer the rim down flat. Repeat in the other sections.

Solder or epoxy the tips of the Frosty Star to the rim of the Sole Sculpture. Little ones are charming on the Christmas tree!

Sole Star Sculpture

Celebration Chandelier

Let's all join together in toasting the A & P for this Celebration Chandelier! You can see they share their savings with their shoppers in unsuspected ways! All-gold, inside and out, ten inches wide and a foot high, these large commercial salad cans are incomparable!

The chandelier is deceptively easy. In fact, it's just a matter of dividing the sides of the can into equal sections, which you then curve with your fingers as the chandelier rests on a tall kitchen stool, and ornamenting

146

the tips with candles or stars. It's not necessary to go to the bother of making Sun Pendants with Sparkler Stars hanging inside them as I did, but they *are* pretty!

Remove the top rim of the can. With your compass, scribe a line around the sides of the can 3″ down from the top (see Cutting Down Cans). Later, you will cut every other arm on the chandelier down to that line.

Divide the sides of the can into thirty-two equal sections, using *shelf* paper or heavy tissue paper, but *not* newspaper, which absorbs the glue and is hard to peel off.

After cutting along the creases, pull the arms out; trim down every other arm; round all the ends; and put awl holes in any that you plan to hang something from. Also strike three pairs of awl holes in the bottom of

Celebration Chandelier

147

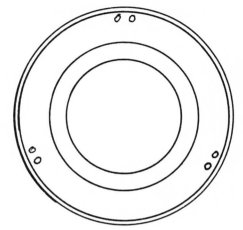

Strike three pairs of awl holes in end of can for wiring chain to chandelier

the can near the rim (see the diagram), from which to suspend the chandelier.

Hammer all the arms until they are perfectly flexible, and shape them with your fingers into easy S curves, resting the chandelier on a high stool as you work around it.

Crimp as many lids (small juice or tomato paste) as you will need for the candles (see Crimping). You *could* use votive candles and save yourself the trouble of gluing on candlesockets, but I think you will find they don't have as attractive a silhouette as the small 3″ Hanukkah candles do. So follow the instructions given for making candlesockets (see Making Scrolls and Candlecups), but make them from smaller strips of tin, ¾″ x 2½″. Glue everything together and if you are going to paint the chandelier, do so now, before adding any pendants.

Sun Pendants and Sparklers

Having used sixteen small juice can lids for the crimped candlecups, I utilized the rest of the cans for the Sun Pendants and the Sparklers, removing the bottom lids for the Sparklers and cutting down the sides of the cans for the Sun Pendants. Should you decide to use Sun Pendants also, divide the cut-down sides into eight equal sections (see Dividing the Sides of Cans into Equal Sections) and simulate flames in each, *free-hand*.

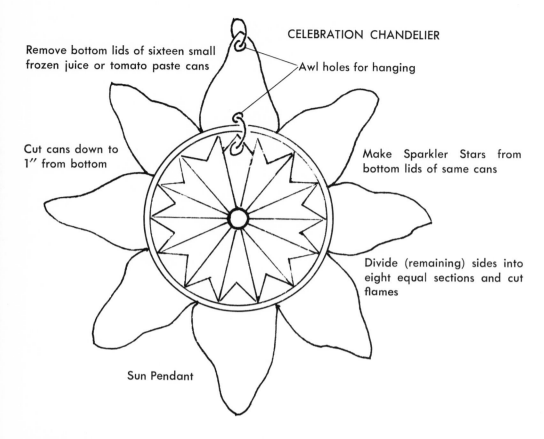

CELEBRATION CHANDELIER

Remove bottom lids of sixteen small frozen juice or tomato paste cans

Awl holes for hanging

Cut cans down to 1″ from bottom

Make Sparkler Stars from bottom lids of same cans

Divide (remaining) sides into eight equal sections and cut flames

Sun Pendant

A little irregularity makes them seem more real. Hammer the sections out to the side; strike awl holes for hanging as indicated; and spray them with Krylon Bright Gold Enamel. (Note: If you have left your chandelier the gold of the can, you would then spray the pendants with Krylon Flat White instead.)

Make the Sparkler Stars on page 57 and spray them with Krylon Flat White or Gold for contrast, depending on the chandelier (see Note above). Link the stars to the pendants (see Making Links) and the pendants to the chandelier.

From a 5′ length of decorator chandelier chain purchased in the hardware store, cut three equal lengths (better count the links to make sure!) and wire them through the pairs of awl holes in the end of the can. Raise your party piece in triumph and suspend it from a hook in the ceiling.

After all that, perhaps you think I should propose a toast to *you! I do!*

SOURCES OF SUPPLY

Allcraft Tool & Supply Company, 20 West 48th Street, New York, New York 10036.

Stocks beautiful ball-peen hammers, Nu-Gold wire, and jewelry findings.

Keeler's Paint Works, Inc., 40 Green Street, New London, Connecticut 06320.

Carries architect's linen along with everything in the way of paints and brushes.

Sy Schweitzer, P. O. Box 106, Harrison, New York 10528.

Also has good selection of jewelry findings, especially chains, plus Epoxy 220. Gives fast service.

Welch's Hardware, 21 Main Street, Westport, Connecticut 06880.

Will supply you with everything you need in the way of essential tools like the Klenk Aviation Snips, the round-nosed pliers, and optional tools like cold chisel, Weldit Cement, etc.

INDEX

(Page numbers in **boldface** are those on which photographs appear)

ABOUT THE AUTHOR

LUCY SARGENT was born in Cambridge, Massachusetts, and grew up in Hingham, Massachusetts, and Syracuse, New York. She was graduated from Wellesley College, where she majored in art, and later worked at the Metropolitan Museum of Art in New York City.

Although marriage and the raising of four children has kept her busy, she has managed to pursue a number of other interests, including gardening, music, writing, and, of course, tincraft.

Lucy Sargent has written two previous books, *Tincraft for Christmas* and *Tincraft*. Her work has been exhibited in New York, Washington, and Berlin, and she has lectured extensively on the subject.

Mrs. Sargent lives in Westport, Connecticut.